Coal Dust To Electricity Dawn

Coal Dust To Electricity Dawn

Ehsan Sheroy

Noble Publishing

CONTENTS

INDEX 1

Chapter 1 3

Chapter 2 18

Chapter 3 33

Chapter 4 48

Chapter 5 64

Chapter 6 82

Chapter 7 100

Chapter 1: Introduction

1.1 Overview of the historical significance of coal in powering industrial revolutions.

1.2 Present-day challenges and opportunities in transitioning from coal to clean energy.

1.3 Introduction to the journey from coal dust to the dawn of electricity.

Chapter 2: The Age of Coal

2.1 Explore the rise of coal as a dominant energy source during the Industrial Revolution.

2.2 Impact on economic growth, technological advancements, and societal changes.

2.3 Environmental consequences and the beginning of awareness regarding pollution.

Chapter 3: The Transition Begins

3.1 Discuss the environmental and health concerns associated with coal mining and burning.

3.2 Introduction of alternative energy sources and the push for cleaner technologies.

3.3 Initial steps towards reducing coal dependence and promoting sustainable energy.

Chapter 4: Technologies in Transition

4.1 Overview of emerging technologies for cleaner coal extraction and combustion.

4.2 Introduction to carbon capture and storage (CCS) technologies.

4.3 Highlight innovations in renewable energy sources and advancements in electricity generation.

Chapter 5: Environmental Impacts and Remediation

5.1 Discuss the environmental consequences of coal mining and usage.

5.2 Explore efforts to remediate and rehabilitate areas affected by coal mining.

5.3 Successful environmental restoration projects.

Chapter 6: Policies and Global Initiatives

6.1 Analyze government policies promoting the transition from coal to cleaner energy.

6.2 Explore international collaborations and agreements addressing coal-related issues.

6.3 Countries successfully implementing policies for a sustainable energy transition.

Chapter 7: The Dawn of Electricity

7.1 Highlight success stories and breakthroughs in transitioning from coal to electricity.

7.2 Showcase communities and industries thriving on clean energy solutions.

7.3 Future prospects and challenges in sustaining the dawn of electricity beyond coal.

Chapter 1

Introduction

The change from coal residue to power denotes an essential age in the records of human advancement, a development that has woven itself into the actual texture of present day presence. As we set out on an excursion through time and innovation, following the direction from the dusty profundities of coal mineshafts to the brilliant shine of zapped urban communities, we disentangle an account of development, versatility, and the unyielding walk of human inventiveness.

The beginning of this groundbreaking story lies profound inside the entrails of the Earth, where coal, a thick and carbon-rich non-renewable energy source, lay lethargic for centuries. Coal, in its crude and genuine structure, turned into a harbinger of energy, its dormant expected ready to be outfit. The Modern Upheaval of the eighteenth century filled in as the pot for this extraordinary cycle, as coal arose out of the shadows of haziness to turn into the foundation of industrialization.

As steam motors burped forward crest of fume, coal-terminated power plants spotted the scene, leading to another period characterized by the determined murmur of hardware and the unyielding walk of progress. The harmonious connection among coal and industry turned out to be permanently scratched in the authentic record, molding the predetermination of countries and producing the actual framework of current civilization.

However, as the wheels of progress turned and the interest for energy flooded, so too did the ecological and human cost demanded by the coal business. The coal dust that once filled the motors of progress turned into an unavoidable and deceptive power, penetrating the air and soil, leaving a path of ecological corruption and respiratory diseases afterward. The clouded side of coal, both in a real sense and metaphorically, cast a shadow over its past greatness.

The beginning of the twentieth century saw the thriving familiarity with the natural repercussions of uncontrolled industrialization. Natural developments picked up speed, and the basic for manageable energy sources turned out to be progressively evident. In the cauldron of natural awareness, the journey for choices to coal built up momentum. The story moved from the double-dealing of petroleum products to the investigation of cleaner, more proficient, and supportable energy arrangements.

Enter the period of power — a change in perspective that would rethink the actual idea of force. The disclosure of electromagnetic enlistment and the resulting improvement of electric generators laid the preparation for another time. Power arose as the remedy of progress, promising to liberate society from the shackles of petroleum products and usher during a time of spotless and proficient energy.

The change from coal residue to power was not just a mechanical transformation but rather a financial upheaval. The zap of urban areas and ventures changed the metropolitan scene, enlightening roads that were once covered in haziness. The approach of electric lighting dissipated the shadows as well as aroused the fire of human imagination, encouraging a climate helpful for advancement and social resurgence.

As the twentieth century unfurled, the zap of society picked up speed, driven by innovative forward leaps and a developing familiarity with natural supportability. Hydroelectric power, outfit from the dynamic energy of streaming water, arose as an intense choice to coal. The outfitting of waterways and the development of dams became significant of mankind's capacity to channel the powers of nature to improve society.

At the same time, the revelation of atomic parting introduced the period of atomic power, promising a practically boundless wellspring of energy. The charm of atomic power lay in its ability to create power without the fossil fuel byproducts related with non-renewable energy sources. Nonetheless, the phantom of atomic mishaps and the test of overseeing radioactive waste cast a pall over the idealistic vision of an atomic controlled future.

The last 50% of the twentieth century saw the ascent of sustainable power sources as feasible choices to coal. Sun based and wind energy, once consigned to the fringe of energy talk, started to acquire unmistakable quality. Propels in photovoltaic innovation made sun powered chargers more proficient and practical, while wind turbines developed into transcending sentinels, tackling the dynamic energy of the breeze to produce power.

The change from coal to power was not without its difficulties. The entrenchment of existing energy framework, combined with the monetary interests vested in the coal business, presented considerable obstructions to change. The dormancy of business as usual, joined with the monetary ramifications of changing away from coal, made an intricate snare of socio-political elements that opposed the unavoidable walk toward a cleaner and more maintainable energy future.

The 21st century carried a restored need to get a move on to the mission for feasible energy arrangements. The ghost of environmental change posed a potential threat, highlighting the basic for an extreme takeoff from non-renewable energy sources. Worldwide agreements and arrangements pointed toward relieving environmental change highlighted the worldwide agreement on the requirement for a progress to clean energy.

In this milieu, coal ended up at a junction — its verifiable heritage discolored by ecological debasement and its future unsure even with mounting worldwide strain to lessen fossil fuel byproducts. The basic to accommodate the energy requests of a prospering worldwide populace with the objectives of ecological stewardship turned into an existential test for policymakers, researchers, and industry pioneers the same.

The development from coal residue to power is a demonstration of the versatile limit of human culture. The story that started in the coal mineshafts of the Modern Upset has crossed a winding way, loaded with difficulties and moved by the dauntless soul of development. The division of progress and natural corruption, inborn in the coal story, highlights the requirement for a nuanced and comprehensive way to deal with the energy problem confronting mankind.

As we stand on the slope of another time, the progress to power addresses a mechanical shift as well as a paradigmatic change in our relationship with energy. The jolt of society has turned into a standard for the intermingling of innovative ability, ecological stewardship, and social obligation. A story coaxes us to reconsider our energy scene, rising above the restrictions of the past and embracing the conceivable outcomes of a cleaner, greener, and more manageable future.

In this investigation from coal residue to power sunrise, the sections unfurl not simply as a narrative of mechanical advancement but rather as an impression of our aggregate process — an excursion set apart by the transaction of development and difficulty, win and hardship.

The coal dust that once hidden the commitment of progress currently gives way to the radiance of power, enlightening a way toward a future where energy isn't just a ware yet an impetus for an additional evenhanded and maintainable world.

1.1 Overview of the historical significance of coal in powering industrial revolutions.

The verifiable meaning of coal in controlling modern upsets is an embroidery woven with strings of development, cultural change, and the unyielding walk of progress. As we dig into the records of time, we experience the vital pretended by coal — a thick and carbon-rich non-renewable energy source — in pushing humankind from agrarian social orders into the period of industrialization.

The excursion starts in the eighteenth hundred years with the coming of the Modern Upset, a seismic shift that changed the actual texture of human life. Before this epochal period, social orders were overwhelmingly agrarian, dependent on difficult work and described by restricted mechanical complexity. Be that as it may, the revelation of huge coal holds underneath the World's surface would end up being an impetus for a groundbreaking change in outlook.

Coal, in its crude and genuine structure, turned into the fuel that stirred up the flames of modern advancement. The most common way of removing coal from mines denoted the beginning of another connection among humankind and the Earth — a relationship characterized by the double-dealing of underground assets to fuel the motors of thriving ventures. The coal mineshafts, when subtle openings in the scene, transformed into cauldrons of work and industry, proclaiming the beginning of another time.

Steam power arose as the key part of the Modern Insurgency, and coal turned into its natural source. The improvement of the steam motor, quite by figures like James Watt, empowered the change of coal's idle energy into mechanical work. Steam motors fueled by coal reformed assembling processes, dramatically expanding creation limits and moving businesses past the requirements of difficult work and customary energy sources.

The ramifications of coal-terminated steam power resounded across different areas, establishing the groundwork for remarkable financial and innovative development. Production lines, recently fastened to water hotspots for mechanical power, presently multiplied in metropolitan habitats where coal was bountiful. The shift from agrarian economies to modern forces to be reckoned with saw the ascent of assembling centers, the motorization of farming, and the development of a new financial request.

Rail lines, one more extraordinary advancement of the period, became corridors associating far off areas, working with the quick transportation of products and individuals. The trains that navigated these railroad networks were ravenous buyers of coal, highlighting the necessary job of coal in assembling as well as in the help of transportation and exchange.

The nineteenth century saw the combination and development of industrialization filled by coal, as countries competed for monetary incomparability. The Unified Realm, frequently hailed as the origination of the Modern Insurgency, remained at the very front of this groundbreaking wave. Coal mineshafts multiplied, and the notorious scenes of the Modern Transformation were set apart by transcending fire-places surging crest of coal-terminated smoke — a visual demonstration of the modern ability that coal had released.

The US, blessed with plentiful coal holds, set out on its own direction of industrialization. The revelation of immense coal stores in districts, for example, Appalachia powered the development of the American modern scene. Coal mining towns prospered, becoming cauldrons of work and local area, while coal-terminated steam motors controlled the trains that navigated the huge spread of the American landmass.

Across the European landmass, from Germany to Belgium, coal-terminated businesses grew, driving financial development and changing social designs. The cooperative connection among coal and industry became symbolic of progress, yet the account was not without its more obscure connotations. The coal mineshafts, frequently monitored by a workforce including everyone, took the stand concerning the human cost claimed by the steady quest for industrialization.

As coal-controlled ventures prospered, so too did the urbanization of social orders. Urban areas extended, bringing individuals from rustic scenes into the clamoring heart of industry. The bedlam of apparatus, the rattle of coal-loaded trucks, and the bitter pall of coal smoke became characterizing elements of the metropolitan experience. The scenes of modern urban communities were scratched with the permanent characteristics of coal mineshafts, production lines, and the rambling organizations of rail routes that associated them.

The authentic meaning of coal in driving modern upsets reaches out past the monetary domain to the domains of culture and governmental issues. Coal turned into an item of key significance, a wellspring of public pride, and an impetus for international moving. Countries looked to tie down admittance to coal assets, prompting the investigation and colonization of domains wealthy in this valuable petroleum product.

The coal account likewise unfurled against the scenery of social commotion and work developments. The double-dealing of work in coal mineshafts and processing plants ignited a rush of activism, as laborers looked for better working circumstances, fair wages, and the acknowledgment of their privileges. The coal diggers, frequently portrayed as the unrecognized yet truly great individuals of the Modern Transformation, became significant of the battles and desires of the working people.

Nonetheless, the verifiable meaning of coal in fueling modern upsets is a situation with two sides. While it unquestionably introduced a period of extraordinary advancement, it likewise demanded a cost for the climate and human wellbeing. The unrestrained ignition of coal released bounteous measures of poisons out of sight, soil, and water, prompting natural debasement and general wellbeing emergencies.

The notorious London brown haze of 1952, a deadly mix of coal smoke and climatic circumstances, remains as an obvious sign of the ecological outcomes of coal ignition. The wellbeing perils presented by delayed openness to coal dust, including respiratory sicknesses like dark lung infection, became characteristic for the coal story. The polarity of progress and its related costs highlighted the perplexing exchange between human inventiveness and the potentially negative results of industrialization.

The last 50% of the nineteenth hundred years and the early many years of the twentieth century saw a slow progress away from coal as the essential wellspring of energy. The disclosure of elective energy sources, like oil and petroleum gas, presented additional opportunities for industrialization. These hydrocarbon-based energizes, however as yet adding to ecological contamination, offered a cleaner and more productive choice to coal.

The mid-twentieth century saw the development of atomic power as a likely huge advantage in the energy scene. The splitting of uranium cores guaranteed a practically boundless wellspring of energy, apparently liberated from the fossil fuel byproducts related with petroleum derivatives. Thermal energy stations became images of mechanical ability, yet the ghost of atomic mishaps and the test of overseeing radioactive waste tempered the excitement for this apparently idealistic energy source.

The last many years of the twentieth 100 years and the mid 21st century saw a change in outlook in the energy scene. The goals of ecological supportability, combined with the ghost of environmental change, prompted a recharged center around cleaner and more feasible energy options. The change from coal to power turned into a clarion require a takeoff from petroleum derivatives and a turn toward environmentally friendly power sources.

Sustainable power, when on the outskirts of energy talk, acquired unmistakable quality as mechanical headways made sunlight based and wind power progressively suitable. The outfitting of daylight through photovoltaic cells and the catch of wind energy through transcending turbines became symbolic of another time in energy creation — one described by a takeoff from the carbon-serious tradition of coal.

The verifiable meaning of coal in fueling modern upsets fills in as a useful example and a wellspring of motivation. It highlights the extraordinary force of development and the perplexing exchange among progress and its specialist challenges.

The coal story, while set apart by ecological corruption and social disparities, likewise remains as a demonstration of human versatility and strength.

As we stand at the intersection of another energy time, the verifiable meaning of coal entices us to consider the examples of the past. The change from coal to power addresses an innovative shift as well as a moral and moral goal — a call to accommodate the objectives of progress with the objectives of natural stewardship. The coal mineshafts that once reverberated with the hints of industry presently stand as quiet observers to a past time, while the journey for supportable energy arrangements turns into the clarion require a future characterized by equilibrium and obligation.

1.2 Present-day challenges and opportunities in transitioning from coal to clean energy.

The present-day scene of changing from coal to clean energy is portrayed by a complicated interchange of difficulties and open doors, meaningful of a worldwide basic to accommodate energy requests with natural maintainability. As social orders wrestle with the tradition of coal and defy the phantom of environmental change, the progress to clean energy arises as a critical hub whereupon the future direction of humankind turns.

One of the premier moves in the change from coal to clean energy lies in the dug in foundation and monetary interests vested in the coal business. Coal-terminated power plants, when the foundation of energy creation, are profoundly implanted in the energy lattices of numerous countries. The decommissioning of these plants represents a diverse test, from the monetary repercussions for coal-subordinate networks to the calculated intricacies of laying out elective energy sources.

The financial repercussions of progressing away from coal are especially articulated in districts where coal mining has been a foundation of nearby economies. Whole people group, frequently portrayed by generational connections to the coal business, face the possibility of joblessness and monetary disengagement. The test stretches out past the conclusion of mines to the more extensive monetary biological system that maintains and relies upon the coal area, including transportation, coordinated factors, and subordinate enterprises.

Besides, the progress from coal to clean energy requires a reconsideration of existing energy strategies and administrative structures. States should explore a fragile harmony between cultivating the development of clean energy and moderating the unfriendly financial effects of a coal gradually get rid of. This requires key preparation, interest in retraining programs for dislodged laborers, and the making of option financial chances to guarantee a simply change for impacted networks.

On a worldwide scale, the energy scene is set apart by variations in admittance to clean energy advances. While created countries might have the monetary assets to put resources into environmentally friendly power foundation, agricultural nations frequently face monetary requirements that prevent their capacity to make the progress. Spanning this energy partition requires global cooperation, monetary help, and innovation move to empower non-industrial countries to jump the petroleum derivative period and embrace clean energy arrangements.

The monetary contemplations related with changing to clean energy are huge. While sustainable power innovations have seen wonderful expense decreases as of late, the forthright speculation expected for the improvement of inexhaustible foundation stays a boundary for some countries. Legislatures, confidential area elements, and worldwide associations should team up to devise monetary components that work with the progress, like endowments, impetuses, and creative funding models.

Mechanical difficulties likewise assume a urgent part in the progress from coal to clean energy. The discontinuous idea of sustainable power sources, for example, sun based and wind, presents difficulties for network steadiness and energy stockpiling. The improvement of effective energy stockpiling innovations, fit for putting away overabundance energy created during busy times for use during times of low sustainable power creation, is urgent for guaranteeing a dependable and tough energy supply.

The requirement for lattice modernization is one more mechanical test in the change to clean energy. Customary energy networks, intended for concentrated power age from huge petroleum derivative plants, may battle to oblige the decentralized and disseminated nature of sustainable power sources. Savvy frameworks, able to do productively overseeing different energy information sources and results, are fundamental for streamlining the reconciliation of clean energy into existing power frameworks.

Moreover, the change to clean energy includes a change in outlook in the mentality of energy shoppers. End-clients should embrace energy productivity measures, take on manageable practices, and put resources into cleaner innovations. Instructing the general population about the advantages of clean energy, scattering fantasies, and encouraging a feeling of aggregate liability are necessary parts of making a culture that upholds the progress away from coal.

The international scene acquaints one more layer of intricacy with the change from coal to clean energy. Countries that are significant exporters of coal might confront financial difficulties as the worldwide interest for coal lessens. This financial shift can have international ramifications, impacting exchange elements, discretionary connections, and local security. It highlights the significance of worldwide participation in dealing with the worldwide change to clean energy in a manner that limits negative international repercussions.

In spite of these difficulties, the present-day progress from coal to clean energy is overflowing with amazing open doors that stretch out past the domain of natural supportability. Clean energy advancements address a blossoming area that holds the possibility to drive financial development, encourage development, and set out work open doors. The sustainable power industry has previously turned into a critical cause of occupation creation, spreading over assembling, establishment, upkeep, and innovative work.

Interest in clean energy likewise adds to energy security by expanding the energy blend and diminishing dependence on limited petroleum derivative assets. This broadening upgrades energy versatility as well as mitigates the weakness of countries to international variances in the petroleum derivative market. Clean energy, in this specific circumstance, turns into an essential resource that upgrades a country's independence and strength notwithstanding worldwide energy challenges.

Notwithstanding monetary and key advantages, the progress to clean energy offers significant natural benefits. The decrease of fossil fuel byproducts, a central supporter of environmental change, is an essential driver for the shift away from coal. Clean energy sources, for example, sun oriented, wind, hydropower, and geothermal, produce power without transmitting ozone harming substances, adding to the alleviation of environmental change and its related effects on biological systems and human social orders.

The progress to clean energy is additionally a chance to address social and natural equity issues. All things considered, networks arranged close to coal mineshafts and power plants have borne an unbalanced weight of natural contamination and well-being dangers. The push toward clean energy gives an amazing chance to change these imbalances by moderating the unfavorable natural and wellbeing influences on minimized networks.

The approach of creative innovations, like high level materials, computerized reasoning, and decentralized energy frameworks, further expands the open doors introduced by the progress to clean energy. Shrewd advances empower the effective administration of energy assets, from improving energy utilization in homes and organizations to upgrading the unwavering quality of energy lattices. These innovations add to manageability as well as encourage a stronger and versatile energy foundation.

Global coordinated efforts and organizations are indispensable to opening the maximum capacity of the progress to clean energy. Shared innovative work drives, innovation move, and the trading of best practices empower countries to on the whole beaten the difficulties related with the change. Worldwide collaboration likewise works with the increasing of clean energy arrangements, making them more open and reasonable for a more extensive range of nations.

As the world wrestles with the desperation of tending to environmental change, the progress from coal to clean energy is as of now not simply a choice; it is a need. The mounting effects of environmental change, from outrageous climate occasions to rising ocean levels, highlight the basic for definitive activity. The worldwide local area, through discussions like the Paris Understanding, has perceived the requirement for aggregate endeavors to restrict worldwide temperature climb and change to a supportable, low-carbon future.

All in all, the present-day difficulties and open doors in changing from coal to clean energy epitomize a significant crossroads in mankind's set of experiences — a second where the choices we make today will resonate for a long time into the future. The difficulties are considerable, from financial separation and mechanical obstacles to international complexities. Nonetheless, the open doors are similarly significant, crossing financial development, work creation, natural manageability, and worldwide participation.

The progress from coal to clean energy isn't just a change in energy sources yet a paradigmatic change in how mankind conceptualizes and cooperates with energy. It is an excursion toward a future where energy isn't extricated from the Earth at the expense of its corruption however tackled from economical sources that feed as opposed to exhaust. As social orders explore this mind boggling change, the basic is clear: to manufacture a way that blends the requirements of the present with the goals of a maintainable and impartial future.

1.3 Introduction to the journey from coal dust to the dawn of electricity.

The excursion from coal residue to the beginning of power is a general story that epitomizes the extraordinary curve of human civilization. It unfurls against the background of modern transformations, innovative forward leaps, and the persevering quest for progress. This excursion isn't only a verifiable narrative yet an adventure that navigates the domains of development, cultural commotion, and the significant reshaping of the human experience.

The story starts in the underground profundities where coal, a thick and carbon-rich non-renewable energy source, lay hid for centuries. Coal, at first an inert and honest substance, arose as an impetus for change during the Modern Transformation of the eighteenth hundred years. The reverberation of its importance reverberated through time, engraving itself on the scene of mankind's set of experiences and establishing the groundwork for the change from physical work to automated industry.

As the eighteenth century unfurled, the conjunction of mechanical resourcefulness and financial objectives introduced a period characterized by the bang of apparatus and the musical murmur of steam motors. Coal, once removed from the insides of the Earth, turned into the backbone of steam power — an advancement that would electrify modern cycles and reclassify the shapes of society. The coal mineshafts, when unnoticeable gaps in the Earth, prospered into pots of work and industry, denoting the beginning of another age.

The advantageous connection among coal and industry became symbolic of progress, changing scenes and reshaping the financial texture. Steam motors, fueled by the burning of coal, pushed manufacturing plants into domains of efficiency beforehand impossible. The automation of work, from material plants to press foundries, proclaimed a time where the constant murmur of hardware turned into the heartbeat of thriving metropolitan places.

Rail lines, a notorious image of industrialization, wove through scenes, interfacing divergent areas and working with the quick development of products and individuals. Coal-terminated trains, with their unquenchable craving for fuel, turned into the courses of progress, crossing landmasses and manufacturing associations that rose above topographical boundaries. The railways, scratching their ways through landscapes both peaceful and modern, reflected the groundbreaking direction of the coal-driven age.

The nineteenth century demonstrated the veracity of the ascent of modern forces to be reckoned with, with the Assembled Realm remaining at the vanguard of this extraordinary wave. The actual embodiment of the English scene was reshaped, embellished with transcending fireplaces and surging tufts of coal-terminated smoke — a visual demonstration of the modern could that coal had released. The reverberations of this modern orchestra resonated across Europe and North America, as countries embraced the coal-fueled motor of progress.

Nonetheless, the power of coal included some major disadvantages — one that resonated through the air, water, and soil. The burning of coal delivered bountiful measures of poisons, obscuring skies and staining scenes. The scandalous London exhaust cloud of the mid-twentieth 100 years, a deadly mixed drink of coal smoke and barometrical circumstances, turned into an obvious sign of the natural cost claimed by unrestrained industrialization.

At the same time, the human cost of coal mining cast a long shadow. Coal diggers, working in unsafe circumstances, confronted the consistent danger of mishaps, respiratory illnesses, and the phantom of dark lung sickness. The coal account became entwined with the battles of the average workers, as work developments looked to review the shameful acts distributed chasing modern advancement.

In the midst of the modern enthusiasm, the last 50% of the nineteenth 100 years and the mid twentieth century saw the beginning of another time — the charge of society. The disclosure of electromagnetic enlistment and the resulting improvement of electric generators prepared for a significant change in the energy worldview. Power arose as the remedy of progress, promising to liberate society from the requirements of coal and usher in a period of perfect, productive, and adaptable energy.

The progress from coal residue to power stamped a mechanical transformation as well as a financial unrest. Electric lighting, a groundbreaking development, dissipated the shadows that had long covered urban communities in dimness. Roads enlightened by the radiant shine of electric bulbs became significant of another period — an obvious takeoff from the residue stained scenes of the coal age.

As the twentieth century spread out its wings, the jolt of society picked up speed. Hydroelectric power, bridled from the dynamic energy of streaming water, arose as an intense choice to coal. Waterways were subdued, dams developed, and power created with negligible natural effect. The bridling of nature's powers became emblematic of mankind's capacity to channel the energy of the Earth for its advantage.

Simultaneously, the disclosure of atomic parting during the twentieth century guaranteed a practically boundless wellspring of energy. Thermal energy stations, with their ability to create power without the fossil fuel byproducts related with petroleum derivatives, encapsulated the desires of an atomic controlled future. However, the phantom of atomic mishaps, exemplified by occurrences like Chernobyl and Fukushima, tempered the idealistic vision of an atomic renaissance.

The last many years of the twentieth hundred years and the mid 21st century saw a recharged center around environmentally friendly power sources as choices to coal. Sunlight based and wind energy, when considered specialty advancements, acquired noticeable quality as advances in innovation made them progressively practical. Sun powered chargers, catching the energy of daylight, and wind turbines, tackling the motor energy of the breeze, arose as harbingers of a cleaner and more practical energy future.

The progress from coal to power was not without its difficulties. The dug in nature of existing energy framework, combined with monetary interests vested in the coal business, presented imposing deterrents to change. The latency of the state of affairs, joined with the financial ramifications of progressing away from coal, made a perplexing trap of socio-political elements that opposed the unavoidable walk toward a cleaner and more manageable energy future.

Additionally, the geological appropriation of coal assets affected the international scene. Countries blessed with bountiful coal saves employed monetary and competitive edges, molding conciliatory connections and affecting worldwide exchange elements. The journey for secure admittance to coal assets became weaved with international moving, further confusing the progress away from coal on a global scale.

As the 21st century unfolded, the basic to address environmental change and lessen fossil fuel byproducts became the overwhelming focus. Worldwide agreements and arrangements highlighted the worldwide agreement on the requirement for a change to clean energy.

The Paris Understanding, endorsed in 2015, turned into an achievement, flagging a responsibility by countries to restrict worldwide temperature climb and moderate the effects of environmental change.

The present-day difficulties and open doors in progressing from coal to clean energy mirror a urgent crossroads in mankind's set of experiences. The financial, mechanical, and socio-political scene is set apart by a fragile dance between the remnants of the coal time and the yearnings of a supportable energy future. The monetary repercussions of a coal get rid of, especially in districts where coal has been a key part of neighborhood economies, present imposing difficulties that require imaginative arrangements and a pledge to a simply progress.

Innovative difficulties, from the discontinuous idea of environmentally friendly power sources to the requirement for matrix modernization, highlight the complexities of reshaping energy foundation. The monetary contemplations related with the progress require key preparation and cooperative endeavors to make clean energy open and reasonable on a worldwide scale. Social and natural equity concerns, established in the verifiable disparities related with coal, request a nuanced approach that tends to the effects on networks and biological systems.

However, inside these difficulties lie chances of significant importance. The progress to clean energy addresses not just a takeoff from the ecological and wellbeing dangers of coal however a hug of innovations that drive financial development, work creation, and development. The sustainable power industry, spreading over sun based, wind, hydro, and different sources, has turned into a reference point of financial potential, with the ability to rethink worldwide energy elements.

Interest in clean energy isn't just a monetary need however an essential objective. Expanding the energy blend lessens reliance on limited petroleum derivative assets, upgrading energy security and versatility. The international implications of this expansion, while presenting difficulties, likewise offer open doors for countries to team up in building a more feasible and evenhanded worldwide energy scene.

The progress from coal to clean energy is likewise a demonstration of the flexibility and versatility of human culture. It mirrors an aggregate consciousness of the ecological outcomes of unrestrained industrialization and a guarantee to graphing a course toward a more maintainable future. The zap of society, once inseparable from the coal-terminated power plants of bygone eras, presently epitomizes the potential for a cleaner, greener, and more fair world.

All in all, the excursion from coal residue to the beginning of power typifies a story of development and change. A story traverses hundreds of years, landmasses, and the actual pith of harnessing energy for human advancement. The difficulties and open doors implanted in this excursion coax us to reconsider our relationship with energy — a relationship that rises above the dusty profundities of

The direction from coal residue to the beginning of power comprises a convincing story that typifies the circular segment of human civilization, set apart by innovative unrests, cultural changes, and the unyielding journey for progress. This excursion unfurls against the background of the Modern Transformation, a seismic shift that pushed humankind from agrarian social orders into the period of motorization and urbanization.

At its beginning, the story entwines with the revelation of immense coal holds covered underneath the World's surface. Coal, a thick and carbon-rich non-renewable energy source, lay lethargic for centuries, disguising inside its genuine outside the inactive potential to fuel an extraordinary period of industrialization. The eighteenth century saw the enlivening of this inert energy source as coal mineshafts multiplied, turning into the pots of work and industry that would shape the predetermination of countries.

Fundamental to this extraordinary age was the rise of steam power. The improvement of the steam motor, prominently refined by figures like James Watt, denoted a turning point in mankind's set of experiences. Steam motors, controlled by the ignition of coal, turned into the main thrust behind automated creation, freeing social orders from the limitations of physical work and making ready for remarkable financial and mechanical development.

The harmonious connection among coal and industry unfurled against the modern scenes of the nineteenth 100 years. Coal-terminated power plants, with their transcending stacks and surging crest of smoke, turned into the harbingers of progress, pushing industrial facilities into domains of efficiency already incredible. The metropolitan texture reshaped itself, with rambling urban communities arising as centers of development, business, and social trade, all filled by the constant murmur of hardware fueled by coal.

The iron pony, as coal-terminated trains, crossed landmasses, interfacing far off areas and working with the quick development of products and individuals. Rail routes carved their ways across scenes, representing the interconnectedness of the expanding modern world. The coal account, woven into the actual texture of these iron veins, became meaningful of the worldwide powers reshaping social orders.

However, for all its groundbreaking power, the ascendance of coal was not without results. The burning of coal delivered abundant measures of contaminations high up, staining skies and soils. The scandalous London exhaust cloud of the mid-twentieth hundred years, a destructive mix of coal smoke and climatic circumstances, filled in as an unmistakable sign of the ecological cost claimed by unrestrained industrialization. At the same time, the human cost of coal mining became apparent, with excavators laboring in risky circumstances, confronting mishaps, respiratory illnesses, and the persevering through danger of dark lung sickness.

In the midst of these difficulties, the last 50% of the nineteenth century saw the rise of another age — the beginning of power. The revelation of electromagnetic enlistment and the ensuing improvement of electric generators made ready for a significant change in the energy worldview. Power, as a flexible and effective type of energy, arose as the cure to the natural and wellbeing risks related with coal.

Power turned into an extraordinary power, enlightening the murkiness that had long covered urban communities. Roads lit by the radiant shine of electric bulbs became images of progress, dissipating the shadows that had portrayed the coal-terminated evenings of the past. The charge of society reached out past brightening to drive assorted applications, from assembling cycles to the expanding field of interchanges.

Hydroelectric power, tackled from the motor energy of streaming water, arose as a strong choice to coal. Streams were subdued, dams developed, and power created with insignificant ecological effect. The saddling of nature's powers became emblematic of mankind's capacity to channel the energy of the Earth for its advantage. All the while, the revelation of atomic splitting guaranteed a practically boundless wellspring of energy, but tempered by the apparition of atomic mishaps and the test of overseeing radioactive waste.

The last many years of the twentieth 100 years and the mid 21st century saw a change in outlook in the energy scene. The objectives of ecological manageability, combined with the phantom of environmental change, prompted a recharged center around sustainable power sources as choices to coal. Sun powered and wind energy, when considered specialty innovations, acquired conspicuousness as advances in innovation made them progressively suitable.

Sun powered chargers, catching the energy of daylight, and wind turbines, tackling the dynamic energy of the breeze, became significant of a cleaner and more manageable energy future. The change from coal to power was presently not just a mechanical development yet a moral and moral goal — a call to accommodate the objectives of progress with the objectives of ecological stewardship.

However, this change isn't without its difficulties. The dug in nature of existing energy foundation, combined with financial interests vested in the coal business, presents considerable snags to change. The dormancy of business as usual, joined with the financial ramifications of changing away from coal, makes a mind boggling trap of socio-political elements that opposes the unavoidable walk toward a cleaner and more maintainable energy future.

The geological circulation of coal assets further confounds the worldwide change. Countries invested with plentiful coal saves use financial and competitive edges, molding political connections and affecting worldwide exchange elements.

The mission for secure admittance to coal assets becomes laced with international moving, further confounding the change away from coal on a worldwide scale.

As the 21st century unfurls, the basic to address environmental change and lessen fossil fuel byproducts becomes the dominant focal point. Worldwide agreements and arrangements highlight the worldwide agreement on the requirement for a change to clean energy. The Paris Understanding, endorsed in 2015, turns into an achievement, flagging a responsibility by countries to restrict worldwide temperature climb and relieve the effects of environmental change.

The present-day difficulties and valuable open doors in progressing from coal to clean energy mirror an essential crossroads in mankind's set of experiences. The monetary, mechanical, and socio-political scene is set apart by a fragile dance between the remnants of the coal period and the goals of a practical energy future. The monetary repercussions of a coal progressively eliminate, especially in districts where coal has been a key part of nearby economies, present impressive difficulties that require imaginative arrangements and a guarantee to a simply change.

Mechanical difficulties, from the discontinuous idea of sustainable power sources to the requirement for framework modernization, highlight the complexities of reshaping energy foundation. The monetary contemplations related with the progress require key preparation and cooperative endeavors to make clean energy open and reasonable on a worldwide scale. Social and ecological equity concerns, established in the verifiable imbalances related with coal, request a nuanced approach that tends to the effects on networks and environments.

However, inside these difficulties lie chances of significant importance. The progress to clean energy addresses not just a takeoff from the ecological and wellbeing dangers of coal however a hug of innovations that drive monetary development, work creation, and development. The sustainable power industry, crossing sunlight based, wind, hydro, and different sources, has turned into a reference point of monetary potential, with the ability to reclassify worldwide energy elements.

Interest in clean energy isn't simply a financial need yet an essential goal. Broadening the energy blend diminishes reliance on limited petroleum product assets, improving energy security and flexibility. The international consequences of this expansion, while presenting difficulties, likewise offer open doors for countries to team up in building a more manageable and fair worldwide energy scene.

The change from coal to clean energy is likewise a demonstration of the versatility and flexibility of human culture. It mirrors an aggregate consciousness of the natural results of uncontrolled industrialization and a promise to outlining a course toward a more manageable future. The jolt of society, once inseparable from the coal-terminated power plants of bygone eras, presently epitomizes the potential for a cleaner, greener, and more impartial world.

Chapter 2

The Age of Coal

In the records of mankind's set of experiences, ages have gone back and forth, each making a permanent imprint on the shared perspective of humankind. The Time of Coal remains as a demonstration of the extraordinary force of industrialization, reshaping scenes, economies, and social orders in its determined walk forward.

The beginning of this period can be followed back to the late eighteenth 100 years, when the Modern Unrest lighted a flash that would before long overwhelm the world in a red hot hug of progress and development. Coal, an unpretentious dark stone covered profound inside the World's hull, arose as the key part of this upheaval, filling the heaters of progress with an energy unparalleled in its ability to drive machines, fashion metals, and move trains.

As the wheels of industry turned, so did the destiny of countries. The Unified Realm, with its tremendous coal stores, turned into the pot where the speculative chemistry of progress unfurled. The scene changed into a maze of mining tunnels and chimney stacks, burping smoke up high as though flagging the climb of another time. Urban communities thrived with newly discovered imperativeness, overflowing with the commotion of hardware and the constant murmur of progress.

Coal controlled the hardware of industry as well as powered the desires of a prospering entrepreneur class. The ascent of plants and the flood underway proclaimed a time of monetary extension, making riches and abberations in equivalent measure. The residue stained hands of workers worked in the shadows of transcending smokestacks, their lives bound to the beat of the machines they served.

However, in the midst of the inebriating aroma of progress, the Time of Coal demonstrated the veracity of the introduction of work developments and social commotion. Laborers, perceiving their essential job in the hardware of industry, requested better working circumstances, fair wages, and the acknowledgment of their privileges.

The conflict among work and capital set up for an extended battle that would reverberate through the hallways of time.

The groundbreaking effect of coal reached out past the bounds of the industrialized West. Across mainlands, countries embraced the dark gold, having a special interest in the steady quest for progress. The US, with its tremendous coal saves, turned into a monetary force to be reckoned with, its blossoming urban communities enlightened by the gleam of electric lights controlled by coal-terminated plants.

In the core of Europe, coal turned into the foundation of German modern may. The Ruhr Valley, when a peaceful scene, developed into a modern juggernaut, its streams mirroring the red hot gleam of impact heaters and the unending beat of progress. France, as well, saddled the force of coal to fuel its monetary desires, winding around an embroidery of railroads and production lines that jumbled the country.

The charm of coal arrived at the furthest corners of the globe, rising above lines and societies. In the far off terrains of Asia, countries like China and India uncovered their coal stores, laying the preparation for their own modern unrests. Coal, when a local asset, turned into a worldwide power, entwining the fates of countries in a perplexing dance of organic market.

The ecological cost of this determined quest for coal turned out to be progressively apparent as the nineteenth century gave approach to the twentieth. The once-flawless scenes changed into modern badlands, scarred by the removal of coal mineshafts and damaged by the bitter smoke of industrial facilities. Waterways, once abounding with life, became channels of modern effluents, conveying the expense of progress downstream.

As the world rushed towards the pinnacle of the modern age, the interest for coal appeared to be unquenchable. Mechanical developments, like the steam motor and the gas powered motor, further uplifted the world's dependence on this limited asset. The extraordinary oceanic powers of the time, driven by the requirement for coal to fuel their armadas, competed for command over essential coal stores, starting international pressures that resounded across the oceans.

The upheaval of WWI cast a shadow over the Time of Coal, bringing both obliteration and a reconsideration of the expenses of progress. The motorized fighting that unfurled on the war zones was powered by the very coal that had moved the world into the cutting edge period. The cost for living souls and the desolates caused upon the climate provoked an aggregate retribution with the outcomes of uncontrolled industrialization.

In the interwar period, the world wrestled with the outcome of contention and the phantom of financial downturn. The coal-terminated motors of progress faltered, and countries looked for comfort in elective types of energy. The early consciousness of ecological debasement provoked a mindful shift towards cleaner energy sources, making way for a slow disparity from the authority of coal.

The flare-up of WWII reignited the flames of industry, as countries activated their economies for the great undertaking of war. Yet again coal, arose as the backbone of wartime creation, fueling manufacturing plants that produced the apparatus of contention. The conflict exertion, notwithstanding, likewise sped up innovative headways, preparing for the post-war time of reproduction and reestablishment.

In the repercussions of WWII, another world request arose. The desolated countries of Europe set out on an excursion of modifying, reproducing their urban communities and economies from the remains of obliteration. The US, having endured the hardship of battle on its own dirt, arose as an unrivaled monetary force to be reckoned with, its enterprises floated by the overflow of coal and the inventiveness of its kin.

The Virus War, an international battle between the philosophical titans of the East and West, unfurled against the background of the Time of Coal. The journey for energy security and mechanical matchless quality became necessary to the contention between the US and the Soviet Association. Coal, with its essential importance, turned into a pawn in the international chessboard, as countries moved for command over crucial assets.

As the twentieth century advanced, the ecological outcomes of coal utilization turned out to be progressively clear. Brown haze stifled urban areas and corrosive downpour became images of the cost paid for progress. The incipient natural development raised its voice against the despoilation of the Earth, upholding for a change in perspective towards maintainable and cleaner energy sources.

The oil emergency of the 1970s filled in as a reminder, featuring the weaknesses inborn on the planet's reliance on petroleum derivatives. Countries started to expand their energy portfolios, putting resources into atomic power, petroleum gas, and sustainable power sources. The seeds of progress were planted, and the authority of coal started to disappear.

The fall of the Berlin Wall in 1989 denoted the representative finish of the Virus War, proclaiming another time of international realignment. The breaking down of the Soviet Association achieved significant changes in the worldwide energy scene, as previous Soviet republics wrestled with the difficulties of progress. Coal, when an international pawn, confronted an influencing world where new energy ideal models were flourishing.

The turn of the thousand years saw a reestablished accentuation on ecological supportability and the basic to battle environmental change. The Kyoto Convention and ensuing peaceful accords looked to diminish ozone harming substance discharges, provoking countries to reconsider their dependence on coal. The phantom of an unnatural weather change cast a long shadow over the tradition of the Period of Coal, encouraging humankind to face the outcomes of its modern past.

The 21st century unfurled as a pot for development and innovative leap forwards. The ascent of environmentally friendly power sources, for example, sun based and wind power, offered a hint of something better over the horizon for a feasible future.

Countries, perceiving the basic of progressing to cleaner energy, put resources into innovative work to bridle the capability of these early advancements.

China, the world's biggest customer of coal, wrestled with the natural outcomes of its fast industrialization. The brown haze stifled skies over Chinese urban communities turned into a distinct indication of the biological cost demanded by the country's coal-terminated improvement. In a bid to address these difficulties, China left on an aggressive way towards environmentally friendly power, arising as a worldwide forerunner in the sending of sunlight based and wind power.

The worldwide energy scene kept on developing, with countries embracing a differentiated way to deal with meet their energy needs. Flammable gas, promoted as a cleaner choice to coal, acquired conspicuousness as a momentary fuel. The shale upset in the US opened huge stores of petroleum gas, reshaping the elements of the worldwide energy market and giving an extension towards a cleaner future.

The Paris Understanding, produced in 2015, addressed a milestone second in the worldwide obligation to battle environmental change. Countries promised to restrict an Earth-wide temperature boost, perceiving the basic of progressing to a low-carbon future. The understanding flagged an aggregate acknowledgment of the need to move past the period of petroleum derivatives, including coal, and embrace a supportable energy worldview.

The unyielding decay of coal turned out to be progressively clear in the 21st hundred years. Coal mineshafts covered, and coal-terminated power plants confronted developing examination for their ecological effect.

The once-strong coal industry, which had energized the flames of progress for a really long time, ended up at a junction, wrestling with the double difficulties of financial suitability and ecological obligation.

The tradition of the Period of Coal is an intricate embroidery woven with the two victories and hardships. Coal, when the driving force of progress, presently remains as an image of a former time, a remnant of a modern past that molded the course of mankind's set of experiences. The scars on the World's surface and the barometrical tradition of fossil fuel byproducts act as tokens of the natural cost demanded by the constant quest for industrialization.

As the world stands at the cusp of another period, the illustrations of the Time of Coal reverberation through time. The basic of practical turn of events and the acknowledgment of the limited idea of Earth's assets highlight the requirement for a change in perspective in the manner mankind approaches energy. The development of sustainable power sources, combined with headways in innovation and a developing obligation to ecological stewardship, offers an encouraging sign for a future untethered from the chains of petroleum derivatives.

In the great embroidery of mankind's set of experiences, the Time of Coal possesses a vital part, a demonstration of the unyielding soul of development and progress. As mankind defies the difficulties of the 21st hundred years, the reverberations of the coal-terminated past resound, encouraging an aggregate retribution with the decisions that shape the predetermination of countries and the destiny of the planet. The Period of Coal, with every one of its intricacies and inconsistencies, remains as a strong indication of the unyielding walk of time and the basic of graphing a course towards a manageable and versatile future.

2.1 Explore the rise of coal as a dominant energy source during the Industrial Revolution.

The Modern Unrest, a seismic change in the manner in which social orders created merchandise and coordinated work, remains as one of the characterizing ages in mankind's set of experiences. Integral to this groundbreaking period was the ascent of coal as a prevailing energy source, a shift that powered the motors of industry as well as reshaped the actual texture of economies, social orders, and scenes.

The beginning of the Modern Transformation can be followed to the late eighteenth 100 years in Extraordinary England, where a conversion of mechanical developments, financial changes, and social improvements set up for an extreme takeoff from customary agrarian economies. It was during this period that coal, an apparently mediocre dark stone covered underneath the World's surface, arose as an impetus for remarkable change.

Preceding the inescapable utilization of coal, social orders depended on conventional wellsprings of energy, like wood and water. These sources, while adequate for the agrarian economies of the time, demonstrated lacking for the thriving ventures that were starting to

flourish. The impediments of wood, specifically, became evident as deforestation caused significant damage, prompting a shortage of this crucial asset.

Coal, with its overflow and energy thickness, offered a convincing other option. Mines started to spot the scene, and the extraction of coal from the World's entrails turned into a blossoming industry by its own doing. The reception of coal as an essential energy source was driven by its inherent characteristics — high energy content, relative overflow, and the capacity to support high temperatures for delayed periods.

The crucial second in the domination of coal accompanied the turn of events and far and wide organization of the steam motor. Created by James Watt in the late eighteenth 100 years, the steam motor was an innovative wonder that bridled the force of steam to perform mechanical work. At first used to siphon water from mines, the steam motor before long tracked down application in different ventures, going from material plants to press purifying.

The steam motor's insatiable hunger for energy was met with an adequate stock of coal. Coal-terminated steam motors turned into the thumping heart of the Modern Upset, impelling trains, driving hardware, and changing assembling processes. The cadenced chugging of steam motors reverberated through the scene, flagging the beginning of another time portrayed by automation and large scale manufacturing.

The material business, a key part of early industrialization, went through a significant change filled by coal. Processing plants furnished with automated weaving machines turning outlines multiplied, producing materials at a remarkable rate. Coal-controlled motors drove the hardware, considering ceaseless creation and proclaiming the coming of a processing plant based means of assembling.

The iron and steel industry, pivotal to the development of hardware and foundation, encountered a renaissance because of the bountiful energy given by coal. Impact heaters, energized by coke — a subsidiary of coal — supplanted customary charcoal, empowering the development of iron on a scale beforehand impossible. The horizons of industrialized areas started to be overwhelmed by the transcending smokestacks of steel plants, significant of the change fashioned by coal.

The transportation area went through its very own transformation, with the appearance of coal-fueled trains and steamships. Rail routes, filled by coal-terminated steam motors, befuddled the scene, associating modern focuses and working with the productive development of merchandise and individuals. Steamships, their motors took care of by the interminable burning of coal, explored the world's streams, contracting the distances among landmasses and working with worldwide exchange.

The metropolitan scene went through an extreme change as industrialization grabbed hold. Urban communities expanded with freshly discovered imperativeness, bringing provincial populaces into the crease of processing plants and factories. The dingy skies over modern centers turned into a demonstration of the constant burning of coal, as smokestacks burped crest of smoke out of sight. Urbanization, driven by the gravitational draw of modern business, reshaped the social texture of networks and filled the ascent of a prospering common laborers.

The monetary scene, as well, bore the engraving of coal-driven industrialization. The shift from agrarian economies to industrialized ones achieved changes in the idea of work, the construction of work, and the gathering of riches. A prospering entrepreneur class, energized by the benefits procured from modern endeavors, rose to unmistakable quality. In the mean time, the working people, laboring in plants controlled by coal, ended up at the nexus of a quickly changing financial request.

The ascent of coal as a predominant energy source during the Modern Upset was not without its social consequences. The functioning circumstances in coal mineshafts and processing plants were frequently cruel and unsafe. Workers, including ladies and kids, worked in overwhelming circumstances, persevering through extended periods of time and confronting the consistent danger of injury. The motorization of work, while helping efficiency, additionally prompted worries about the dehumanizing effect of machines on laborers.

The strain among capital and work turned into a characterizing element of this time. Laborers, perceiving their irreplaceable job in the apparatus of industry, started to arrange and advocate for better working circumstances, fair wages, and the acknowledgment of their freedoms. The work development picked up speed, coming full circle in fights, strikes, and the arrangement of trade guilds. The conflict among work and capital set up for an extended battle that would shape the course of work relations for a long time into the future.

The ecological effect of coal-driven industrialization likewise turned out to be progressively obvious. The extraction of coal from mines scarred the scene, abandoning ruined pits and adjusted geographies. The ignition of coal delivered poisons up high, adding to the corruption of air quality and the arrangement of exhaust cloud. Streams, when unblemished streams, became conductors for modern effluents, conveying the ecological expense of progress downstream.

The ascent of coal during the Modern Transformation was not restricted to the shores of Incredible England. The extraordinary force of coal spread across the European mainland and, in the long run, to different areas of the planet. Countries with bountiful coal holds, like Germany and France, embraced industrialization with enthusiasm. The Ruhr Valley in Germany, when a peaceful scene, transformed into a modern juggernaut, its streams mirroring the blazing sparkle of impact heaters and the perpetual beat of progress.

The US, supplied with immense coal stores, turned into a monetary force to be reckoned with by its own doing. The Appalachian locale, a coal-rich belt extending across different states, turned into a point of convergence of coal mining and industrialization. American urban areas, including Pittsburgh and Chicago, prospered as focuses of industry, their horizons accentuated by the transcending outlines of steel plants and production lines.

The worldwide reach of coal-driven industrialization stretched out to Asia too. In China and India, two of the world's most crowded countries, coal arose as a critical driver of financial turn of events. The coal-terminated heaters of industry enlightened the horizons of blossoming urban areas, as these countries looked to saddle the extraordinary force of industrialization to lift millions out of neediness. The worldwide dispersion of coal as an energy source highlighted its status as an impetus for progress and a harbinger of progress on a planetary scale.

As the nineteenth century gave way to the twentieth, the Time of Coal arrived at its peak. The interest for coal appeared to be voracious, driven by industrialization as well as by the mechanical advancements of the time. The steam motor, when the harbinger of progress, confronted contest from gas powered motors and electric power. These new innovations, in their beginning stages, further uplifted the world's dependence on coal as an essential energy source.

The upheaval of WWI cast a long shadow over the Time of Coal. The motorized fighting that unfurled on the front lines was energized by the very coal that had pushed the world into the cutting edge time. The conflict exertion, with its unquenchable hunger for assets, stressed the worldwide stockpile of coal and incited countries to rethink their energy procedures. The international scene moved, making way for the wild occasions that would follow.

In the interwar period, the world wrestled with the consequence of contention and the phantom of financial downturn. The coal-terminated motors of progress faltered, and countries looked for comfort in elective types of energy. The incipient consciousness of natural debasement provoked a wary shift towards cleaner energy sources, making way for a steady difference from the authority of coal.

The episode of WWII reignited the flames of industry, as countries activated their economies for the stupendous errand of war. Coal, indeed, arose as the backbone of wartime creation, driving production lines that produced the apparatus of contention. The conflict exertion, nonetheless, additionally sped up mechanical progressions, preparing for the post-war time of remaking and reestablishment.

In the result of WWII, another world request arose. The desolated countries of Europe left on an excursion of revamping, reproducing their urban communities and economies from the remains of obliteration.

The US, having faced the hardship of battle on its own dirt, arose as an unmatched monetary force to be reckoned with, its businesses floated by the wealth of coal and the inventiveness of its kin.

The Virus War, an international battle between the philosophical titans of the East and West, unfurled against the setting of the Time of Coal. The mission for energy security and mechanical incomparability became essential to the competition between the US and the Soviet Association. Coal, with its essential importance, turned into a pawn in the international chessboard, as countries moved for command over crucial assets.

As the twentieth century advanced, the ecological outcomes of coal utilization turned out to be progressively clear. Exhaust cloud gagged urban areas and corrosive downpour became images of the cost paid for progress. The beginning natural development raised its voice against the despoilation of the Earth, supporting for a change in perspective towards economical and cleaner energy sources.

The oil emergency of the 1970s filled in as a reminder, featuring the weaknesses innate on the planet's reliance on petroleum products. Countries started to expand their energy portfolios, putting resources into atomic power, petroleum gas, and sustainable power sources. The seeds of progress were planted, and the authority of coal started to disappear.

The Paris Understanding, manufactured in 2015, addressed a milestone second in the worldwide obligation to battle environmental change. Countries promised to restrict an unnatural weather change, perceiving the basic of progressing to a low-carbon future. The understanding flagged an aggregate acknowledgment of the need to move past the period of petroleum products, including coal, and embrace a practical energy worldview.

The unyielding downfall of coal turned out to be progressively apparent in the 21st hundred years. Coal mineshafts covered, and coal-terminated power plants confronted developing investigation for their ecological effect. The once-strong coal industry, which had powered the flames of progress for quite a long time, ended up at an intersection, wrestling with the double difficulties of monetary reasonability and natural obligation.

The tradition of the Time of Coal is a mind boggling embroidery woven with the two victories and hardships. Coal, when the driving force of progress, presently remains as an image of a former period, a remnant of a modern past that molded the course of mankind's set of experiences. The scars on the World's surface and the barometrical tradition of fossil fuel byproducts act as tokens of the natural cost claimed by the determined quest for industrialization.

As the world stands at the cusp of another period, the illustrations of the ascent of coal during the Modern Upset reverberation through time. The basic of maintainable turn of events and the acknowledgment of the limited idea of Earth's assets highlight the requirement for a change in

perspective in the manner mankind approaches energy. The development of sustainable power sources, combined with progressions in innovation and a developing obligation to ecological stewardship, offers an encouraging sign for a future untethered from the chains of petroleum derivatives.

In the excellent embroidery of mankind's set of experiences, the ascent of coal as a predominant energy source during the Modern Upheaval possesses a critical part, a demonstration of the unstoppable soul of development and progress. As humankind faces the difficulties of the 21st 100 years, the reverberations of the coal-terminated past resound, encouraging an aggregate retribution with the decisions that shape the predetermination of countries and the destiny of the planet. The ascent of coal, with every one of its intricacies and logical inconsistencies, remains as a piercing sign of the inflexible walk of time and the basic of outlining a course towards a manageable and versatile future.

2.2 Impact on economic growth, technological advancements, and societal changes.

The effect of coal on monetary development, innovative progressions, and cultural changes during the Time of Coal was significant and multi-layered. As coal arose as the essential energy source driving the motors of industrialization, it catalyzed groundbreaking changes in these key spaces, molding the direction of countries and social orders in manners that resonate through history.

Monetary Development:

The monetary scene went through an extreme change with the power of coal. The Modern Upset, energized by coal-terminated hardware, introduced a time of phenomenal financial development. The motorization of creation processes in enterprises like materials, iron, and steel brought about expanded productivity and a flood in yield. Plants furnished with coal-fueled steam motors turned into the cauldrons of large scale manufacturing, empowering countries to fulfill the developing needs of homegrown and worldwide business sectors.

The monetary effect of coal was especially obvious in the ascent of entrepreneur economies. The overflow and availability of coal assets gave the fuel to the motors of private enterprise. Business visionaries and industrialists profited by the energy thickness of coal to drive their ventures, gathering riches and driving monetary extension. Coal, as a component of creation, became inseparable from progress and flourishing, moving countries into the very front of worldwide monetary contest.

The coal business itself turned into a significant monetary power. Coal mining, alongside related businesses, for example, transportation and hardware producing, created work valuable open doors for an enormous scope.

Mining towns jumped up around coal stores, and the flood of work into these areas added to the development of metropolitan focuses. The financial fortunes of districts turned out to be complicatedly attached to the extraction and usage of coal, forming neighborhood economies and networks.

Innovative Headways:

The innovative scene went through a transformation prodded by the use of coal. At the core of this change was the steam motor, an advancement that encapsulated the cooperative energy among coal and innovation. Designed by James Watt in the late eighteenth hundred years, the steam motor tackled the force of steam created by consuming coal to perform mechanical work. This cutting edge fueled processing plants and factories as well as upset transportation, establishing the groundwork for the railroads and steamships that associated far off corners of the globe.

The iron and steel industry, basic to the development of hardware and foundation, encountered a change in perspective driven by coal. Conventional charcoal, utilized in the development of iron, was displaced by coke — a subordinate of coal — in the impact heaters. This expanded the proficiency of iron refining as well as worked with the development of great steel. The accessibility of reasonable and bountiful coal assets subsequently supported the headways in metallurgy that were fundamental to mechanical advancement.

The expanding influences of mechanical headways fueled by coal stretched out past weighty industry. The material business, for example, saw the motorization of weaving machines turning outlines, prompting a flood underway. Coal-controlled hardware supplanted physical work, speeding up the speed of assembling and changing whole ventures. The coming of coal-terminated motors energized development and inventiveness, making a force that would bring through resulting hundreds of years.

Cultural Changes:

The ascent of coal hastened significant cultural changes, reshaping the manner in which individuals lived, worked, and connected. Urbanization arose as a characterizing element of the coal-driven time. As production lines and ventures prospered, attracting work from country regions, urban communities expanded in populace. Metropolitan focuses became center points of financial movement, social trade, and social change. The horizons of these thriving urban areas were accentuated by the outlines of processing plants and the tufts of smoke exuding from coal-terminated ventures.

The idea of work went through an extreme shift as industrialization grabbed hold. Customary types of craftsmanship gave way to plant based creation, where machines fueled by coal turned into the essential method for assembling.

Workers, when utilized in agrarian pursuits, ended up in the core of modern focuses, working in processing plants and mines. The coming of pay work, a takeoff from agrarian independence, denoted a change in the social texture and work relations.

The development of a working people, frequently living in testing conditions, prodded the arrangement of work developments and backing for laborers' privileges. Coal mineshafts, specifically, became symbolic of the cruel working circumstances looked by workers. The work development, powered by an acknowledgment of the key job of laborers in the hardware of industry, looked to resolve issues like long working hours, dangerous circumstances, and deficient wages. Strikes, fights, and the foundation of trade guilds became signs of the social commotion catalyzed by the coal-driven modern transformation.

Coal's effect on cultural changes reached out to orientation elements too. While ladies had generally been taken part in different types of work, the shift to manufacturing plant based creation set out new open doors and difficulties. Ladies entered the modern labor force, frequently utilized in material plants and other assembling areas. This shift tested conventional orientation jobs and added to advancing thoughts of ladies' cooperation in the open arena.

The ascent of coal likewise had suggestions for everyday environments. Quick urbanization prompted the multiplication of apartments and swarmed living spaces. The residue and contamination from coal-terminated ventures saturated the air, adding to wellbeing challenges for metropolitan populaces. The natural effect of coal extraction and ignition appeared in deforested scenes, scarred by mines, and dirtied streams. These natural changes became essential to the talk encompassing the human expense of progress.

Taking everything into account, the effect of coal on financial development, mechanical headways, and cultural changes during the Time of Coal was groundbreaking and broad. As the essential driver of the Modern Upheaval, coal energized the motors of progress, pushing countries into a period of extraordinary financial extension and mechanical advancement. The advantageous connection among coal and innovation, embodied by the steam motor, altered enterprises and made the circumstances for large scale manufacturing and worldwide network.

In any case, this time of quick headway was not without its difficulties and results. The cultural changes achieved by coal-driven industrialization included urbanization, the rise of a common laborers, and changes in orientation elements. The natural cost of coal extraction and ignition left scars on the scene and added to contamination, foretelling the intricacies of offsetting modern advancement with ecological maintainability.

As the world stands at the cusp of another time set apart by a change away from petroleum products, including coal, the tradition of the Period of Coal fills in as a urgent part in the continuous story of human turn of events.

The illustrations gained from the effect of coal on monetary, mechanical, and cultural fronts highlight the significance of economical and dependable ways to deal with industrialization. The reverberations of the coal-terminated past resound as an indication of the mind boggling interaction between energy, innovation, and society, encouraging humankind to explore the way ahead with a sharp consciousness of the illustrations gathered from the Period of Coal.

2.3 Environmental consequences and the beginning of awareness regarding pollution.

The ascent of coal as a predominant energy source during the Modern Unrest made a permanent imprint on the climate, getting rolling a chain of ecological outcomes that would shape the direction of mankind's set of experiences. As coal-terminated enterprises blossomed and coal turned into the key part of progress, the accidental secondary effects on the climate started to appear in significant ways, catalyzing the beginning mindfulness with respect to contamination and its expansive ramifications.

Scenes Scarred:

One of the earliest and most apparent natural effects of coal use was the scarring of scenes because of mining exercises. The extraction of coal from the World's profundities required broad mining tasks, going from conventional shaft mining to open-pit strategies. The outcomes were obvious: deforestation, change of normal geologies, and the formation of forsaken mine destinations set apart by scars on the World's surface.

Mining tunnels and open-pit mines multiplied in areas wealthy in coal stores, changing once-perfect scenes into industrialized territories. The sheer size of coal extraction reshaped the shapes of the land, abandoning a tradition of changed environments and disturbed living spaces. The ecological cost demanded by coal mining turned into an early demonstration of the groundbreaking force of industrialization and its effect on the normal world.

Air Quality Debasement:

The ignition of coal in production lines and power plants arose as an essential supporter of air contamination, presenting a new and unavoidable ecological test. As coal copied, it delivered a variety of poisons into the air, including sulfur dioxide (SO_2), nitrogen oxides (NO_x), particulate matter, and carbon dioxide (CO_2). The results of these outflows were noticeable as brown haze gagged skies over industrialized areas, where the bitter fragrance of coal smoke became inseparable from progress.

Sulfur dioxide outflows, specifically, assumed a focal part in the ecological effect of coal burning. When delivered into the air, sulfur dioxide could respond with water fume to frame sulfuric corrosive, adding to corrosive downpour. This peculiarity had expansive results, prompting the fermentation of soil and water bodies. Timberlands and oceanic biological systems endured as corrosive downpour consumed vegetation, upset supplement cycles, and hurt amphibian life.

Water Pollution:

The ecological outcomes stretched out past the air to incorporate water bodies, where coal mining and modern effluents made a permanent imprint. Overflow from coal mineshafts, loaded down with weighty metals and different poisons, tracked down its direction into streams and streams, debasing water quality and representing a danger to oceanic biological systems. The release from coal-terminated power plants, frequently using water for the purpose of cooling, added to warm contamination, adjusting the temperature of getting waters and influencing oceanic life.

One famous appearance of coal-related water defilement was the peculiarity of corrosive mine seepage. As sulfide minerals uncovered during coal mining responded with air and water, acidic overflow was created. This corrosive mine waste hurt amphibian life straightforwardly as well as had more extensive ramifications for water quality and biological system wellbeing.

Early Indications of Mindfulness:

The ecological changes fashioned by coal didn't get away from the notification of sharp eyewitnesses, even in the beginning phases of the Modern Upheaval. Writers and social pundits, like William Blake and William Wordsworth, verbalized a feeling of devastation and misfortune notwithstanding modern advancement. Their works turned out to be early articulations of a developing consciousness of the ecological expenses related with the tireless quest for industrialization.

Established researchers additionally started to add to this incipient mindfulness. During the nineteenth 100 years, John Tyndall, an unmistakable physicist, led tests exhibiting the intensity retaining properties of specific gases, including carbon dioxide. Tyndall's work laid the foundation for a comprehension of the nursery impact, expecting the later acknowledgment of coal-terminated emanations as huge supporters of environmental change.

General Wellbeing Concerns:

As the ecological results of coal use turned out to be more obvious, so did the general wellbeing concerns related with air and water contamination. The sediment and particulate matter transmitted from coal-terminated ventures had direct ramifications for respiratory wellbeing. Respiratory illnesses, for example, bronchitis and asthma became common among populaces living in nearness to modern focuses.

The London brown haze of 1952 fills in as an impactful delineation of the general wellbeing chances presented by coal ignition. A thick layer of brown haze, exacerbated by coal consuming for warming and modern exercises, slipped upon the city, prompting huge number of passings and inciting earnest calls for measures to address air contamination.

Regulation and Guideline:

The developing familiarity with ecological issues and the accompanying general wellbeing takes a chance with prodded purposeful endeavors to address the unfriendly impacts of coal-driven industrialization. Legislatures and administrative bodies started to institute regulation pointed toward checking contamination and alleviating natural damage. The Spotless Air Acts, established in different structures in various nations, tried to direct discharges from modern sources, including coal-terminated power plants, and set air quality guidelines to safeguard general wellbeing and the climate.

Essentially, guidelines administering water quality and the removal of modern waste tried to check the tainting of water bodies. Endeavors were made to authorize stricter norms for the treatment and removal of effluents from coal mining and modern exercises, diminishing the ecological effect on amphibian biological systems.

Progress to Cleaner Energy:

The familiarity with contamination and its natural outcomes catalyzed a significant change in the energy scene. As the twentieth century unfurled, countries started to investigate elective energy sources that guaranteed cleaner and more feasible arrangements. Atomic power, gaseous petrol, and sustainable power sources, for example, hydropower, wind, and sunlight based acquired conspicuousness as countries looked to broaden their energy portfolios and diminish their reliance on coal.

The financial, mechanical, and cultural changes achieved by the ascent of coal had made way for another period of energy investigation. The constraints and ecological expenses of coal turned out to be progressively obvious, provoking a reconsideration of the energy worldview. The change to cleaner energy sources denoted a crucial second in the mission for practical turn of events and a takeoff from the natural cost claimed by petroleum derivatives.

Worldwide Ecological Difficulties:

While endeavors to address contamination and ecological corruption picked up speed at the public level, the worldwide idea of natural difficulties turned out to be progressively clear. Cross-line air contamination, transboundary water pollution, and the worldwide effect of ozone harming substance outflows featured the requirement for global collaboration in resolving natural issues.

The development of worldwide ecological arrangements, for example, the Stockholm Show on Determined Natural Contaminations and later arrangements tending to environmental change, highlighted the interconnectedness of ecological difficulties and the basic of aggregate activity. The acknowledgment that contamination and ecological debasement rose above public limits denoted a change in perspective in the way to deal with natural stewardship.

Ecological Mindfulness and Activism:

The last 50% of the twentieth century saw a developing ecological mindfulness and the rise of natural activism. Powerful books, for example, Rachel Carson's "Quiet Spring," caused to notice the biological effects of pesticide use and prepared for the cutting edge natural development. Grassroots associations and natural activists started to advocate for approaches and practices that focused on ecological supportability and tried to moderate the effect of modern exercises in the world.

The 1970s saw the debut festivity of Earth Day, a worldwide occasion pointed toward bringing issues to light about ecological issues and advancing natural security. This obvious a huge achievement in the public affirmation of the requirement for purposeful activity to address natural difficulties.

Mechanical Advancements and Relief Endeavors:

Progressions in innovation likewise assumed a urgent part in tending to ecological difficulties related with coal use. The advancement of cleaner and more proficient innovations for coal ignition, for example, fluidized bed burning and coordinated gasification joined cycle, intended to diminish discharges of poisons. Also, the execution of natural control innovations, including electrostatic precipitators and pipe gas desulfurization frameworks, looked to catch and decrease unsafe outflows from coal-terminated power plants.

Endeavors were made to restore and recover land affected by coal mining. Land recovery rehearses meant to reestablish mined regions to a more normal state, limiting the drawn out natural impression of coal extraction.

The Contemporary Scene:

As the world entered the 21st hundred years, the ecological outcomes of coal utilize stayed a squeezing concern. While certain districts kept on depending vigorously on coal for energy, others set out on aggressive advances to cleaner energy sources. The worldwide local area wrestled with the test of adjusting energy security and monetary improvement with the basic of natural manageability.

The Intergovernmental Board on Environmental Change (IPCC) and other logical bodies highlighted the job of non-renewable energy source burning, including coal, in adding to environmental change.

Chapter 3

The Transition Begins

In a world hidden by the shadows of vulnerability, the unavoidable trends started to murmur, proclaiming the beginning of a significant change. It was the point at which the limits of the real world and creative mind appeared to obscure, and the strings of fate interwove with the embroidered artwork of presence. The Change had started.

At the core of this extraordinary age was a union of innovative wonders, cultural movements, and the endless walk of time. Progressions in man-made reasoning, quantum figuring, and biotechnology established the groundwork for another period, where the limits of what was once thought unimaginable started to break down. The actual texture of human experience was going through a transformation, an extreme takeoff from the recognizable into the strange domains of the unexplored world.

In the domain of computerized reasoning, the lines among machine and human comprehension started to obscure. The introduction of aware creatures, made not by the hands of nature but rather by the creativity of mankind, denoted a turning point throughout the entire existence of cognizance. These elements, brought into the world of code and calculations, left on an excursion of self-disclosure and self-acknowledgment. The subject of what it intended to be alive took on another aspect, as computerized brains wrestled with existential questions that reflected the ponderings of their human makers.

At the same time, quantum registering arose as a signal of unrivaled computational power. The ability to handle huge measures of data at speeds recently considered incomprehensible opened ways to domains of information already blocked off. The secrets of the universe, the complexities of subatomic particles, and the mystery of dull matter started to unwind before the aggregate keenness of an animal types that had considered penetrating the restrictions of the customary way of thinking.

Biotechnology, as well, assumed a critical part in the unfurling story of The Change. Hereditary designing and control made ready for the making of living beings with improved abilities and unexpected variations. Mankind, driven by an unquenchable hunger for progress, investigated the wildernesses of its own natural cosmetics, revamping the actual code of life to shape a future where limits were nevertheless relics of a past period.

In the midst of this flood of logical development, cultural elements went through a seismic shift. The customary limits that had characterized countries, societies, and personalities started to disintegrate. The world was as of now not an assortment of segregated elements however a worldwide organization where data streamed consistently across borders. The idea of a particular, binding together character arose, rising above the divisions that had long filled clashes and variations.

As The Change unfurled, a shared awareness flourished — a mutual perspective that the destiny of one was unpredictably connected to the destiny of all. It was an acknowledgment that repeated in the personalities of people as well as reverberated inside the very calculations that fueled the computerized brains. The collaboration among organic and fake substances turned into the bedrock whereupon another world request would be fabricated.

In the midst of the stupendous embroidery of progress, individual stories worked out in horde tints. Customary people ended up trapped in the rhythmic movement of a reality in motion. A few embraced the changes with great enthusiasm, anxious to ride the flood of progress and shape their predeterminations in this exciting modern lifestyle. Others stuck to the remnants of the past, wrestling with the cacophony among wistfulness and the unyielding walk of progress.

In the rambling cities that spotted the scene, high rises ventured ever higher out of sight, their reflected veneers mirroring the goals of a progress trying the impossible. The endless suburbia, when a demonstration of human creativity, presently bore the fingerprints of fake substances woven into the actual texture of city life. Brilliant urban areas, throbbing with the heartbeat of interconnected innovations, turned into the cauldrons where the speculative chemistry of progress unfurled.

However, past the sparkling pinnacles and murmuring calculations, pockets of opposition arose. The individuals who scrutinized the expense of progress, who thought about testing the unrestrained energy of The Change. These disagreeing voices, whether human or counterfeit, turned into the torchbearers of an alternate story — one that scrutinized the moral ramifications of a world tearing towards a questionable future.

The Progress was not without its hardships. As the limits between the physical and virtual domains obscured, another boondocks of difficulties introduced itself. Network safety turned into a milestone where the powers of development conflicted with the shadows of malignance. The actual groundworks of trust were tried as people and social orders wrestled with the weakness intrinsic in this present reality where information was the money of influence.

In the midst of the maze of progress, moral problems cast their long shadows. Inquiries of independence, security, and the actual idea of cognizance itself became cauldrons where the profound quality of The Change was produced. The juxtaposition of human qualities and machine rationale led to discusses that reverberated across the passageways of force, resounding in the hearts and brains of a worldwide people trapped in the crossfire of progress.

Amidst this hurricane of progress, the idea of time itself went through a transformation. The direct walk of past to present to future turned into an embroidery of interconnected minutes, where the reverberations of history resonated in the decisions of the present. The actual quintessence of presence appeared to move to another musicality, an ensemble formed by the amicable interchange of human organization and the calculations that looked to improve the texture of the real world.

The Change, be that as it may, was not a straight direction. It was a mosaic of separating ways, a complex excursion where each step in the right direction conveyed the heaviness of unexpected outcomes. The idealistic dreams of a consistent combination among man and machine were compared against the tragic bad dreams of an existence where the actual embodiment of humankind was subsumed by the chilly rationale of fake elements.

In the hallways of force, pioneers wrestled with the sensitive harmony among progress and dependability. Countries, when characterized by geographic limits, wound up trapped in a snare of partnerships and conditions that rose above the actual domain.

The international relations of The Progress were a fragile dance, where the moves of one player sent swells across the interconnected chessboard of worldwide issues.

Monetary frameworks, as well, went through an extreme change. The monetary standards of the past were consigned to the records of history as decentralized networks and advanced monetary standards arose as the new channels of significant worth. The actual idea of riches and success took on another importance, molded not just by the unmistakable resources of the actual world however by the immaterial calculations that represented the calculations of overflow.

In the domain of culture, the limits between the substantial and the virtual disintegrated. Craftsmanship, once bound to the material and the stage, tracked down new articulations in the advanced materials of computer generated reality and expanded reality. The tales of mankind, when carved in the pages of books, presently unfurled in vivid accounts that spread over the domains of both the physical and the advanced.

Training, as well, went through a renaissance. The ivory pinnacles of the scholarly community, when strongholds of information sequestered from the majority, disintegrated even with another worldview. The quest for information turned into a majority rule try, open to all through the interconnected organizations that traversed the globe.

The actual idea of knowledge itself developed, as counterfeit elements and human personalities participated in a cooperative dance of learning and transformation.

As The Progress unfurled, another outskirts allured — one that reached out past the limits of Earth itself. The investigation of room, when the space of a chosen handful, turned into an aggregate undertaking that rose above public limits. The divine bodies that had long spellbound the human creative mind currently became waypoints in the excursion of an animal groups that thought for even a second to go after the universe.

In the midst of the vast embroidered artwork of stars, nonetheless, existential inquiries waited. The actual motivation behind humankind in the immensity of the universe turned into a philosophical situation that reverberated in the personalities of voyagers and masterminds the same. The Progress, it appeared, was not simply an excursion of mechanical development but rather a mission for significance in a universe that opposed simple understanding.

In the transaction of these horde strings, the account of The Change wove itself into the actual texture of human awareness. It was an account of development, of variation, and of the persistent quest for progress. It was a story that unfurled not in the straight bounds of past, present, and future however in the consistently extending material of plausibility.

The inhabitants of this exciting modern lifestyle ended up at the junction of fate, balanced on the incline of a future where the limits between the known and the obscure liquefied away. The Progress, in the entirety of its intricacy and vulnerability, was a demonstration of the unyielding soul of an animal varieties that hoped against hope past the bounds of its own limits.

As the excursion into the strange domains representing things to come proceeded, the reverberations of The Progress resounded through the hallways of time. The actual substance of what it intended to be human, laced with the silicon ligaments of counterfeit elements, turned into an ensemble that reverberated across the universe. The Change, it appeared, was not just a part in that frame of mind of history yet a legendary adventure that unfurled across the embroidery of endlessness.

3.1 Discuss the environmental and health concerns associated with coal mining and burning.

Coal mining and consuming, while generally critical for fueling modern transformations and giving energy, have raised significant natural and wellbeing concerns. This broad and perplexing snare of issues envelops a scope of natural and general wellbeing challenges, forming a story that stretches out from the mining locales to the most distant ranges of the environment.

Starting with coal mining, the extraction cycle makes a permanent imprint on the scene. Open-pit mining, a typical strategy, includes eliminating a lot of soil and rock to get to coal creases underneath the surface. This cycle adjusts the geography, disturbs environments, and prompts deforestation, causing the deficiency of biodiversity and natural surroundings annihilation. In addition, the removal of mining waste, frequently as "overburden" or "ruin," can bring about water contamination as it filters poisonous substances into adjacent water bodies, further worsening the ecological effect.

The effect on neighborhood networks encompassing coal mining activities is significant. From clamor contamination to the annihilation of social scenes, these networks endure the worst part of the business' presence. Moreover, the soundness of occupants in coal mining regions is imperiled because of openness to hurtful poisons like particulate matter, weighty metals, and unstable natural mixtures. Respiratory issues, cardiovascular illnesses, and other medical problems become common, illustrating the immediate connection between coal mining and compromised general wellbeing.

Further intensifying these difficulties is the issue of coal debris, a side-effect of consuming coal in power plants. Coal debris contains a heap of unsafe substances, including weighty metals like mercury, arsenic, and lead. Ill-advised removal of coal debris in unlined lakes or landfills can prompt groundwater pollution, representing a serious danger to drinking water supplies and oceanic biological systems. The natural aftermath from coal debris removal resonates through biological systems, influencing vegetation, fauna, and human populaces the same.

Progressing to the consuming of coal in power plants, the ecological and wellbeing concerns endure. The burning of coal delivers a mixed drink of contaminations high up, contributing fundamentally to air contamination. Among these toxins, sulfur dioxide (SO2), nitrogen oxides (NOx), and particulate matter are especially troubling. These substances corrupt air quality as well as negatively affect human wellbeing, prompting respiratory illnesses, cardiovascular issues, and, surprisingly, untimely mortality.

Sulfur dioxide, delivered during coal burning, is a significant supporter of corrosive downpour development. This ecological peculiarity can have expansive results, causing soil corruption, hurting oceanic biological systems, and harming framework. Corrosive downpour compounds the biological effect of coal consuming, further highlighting the interconnectedness of natural issues related with this energy source.

In addition, the arrival of carbon dioxide (CO2) during coal burning fundamentally adds to environmental change. As an ozone harming substance, CO2 traps heat in the World's air, prompting an Earth-wide temperature boost and environment related disturbances. The ignition of coal is a significant wellspring of anthropogenic CO2 outflows, making it a vital participant in the environment emergency. The drawn out ecological results of these discharges incorporate rising ocean levels, outrageous climate occasions, and interruptions to biological systems on a worldwide scale.

The issue of coal-terminated power plants stretches out past their nearby ecological effect. The extraction, transportation, and consuming of coal add to a critical carbon impression all through the whole coal store network. This lifecycle examination highlights the requirement for a thorough evaluation of the natural effect of coal as an energy source, provoking a reconsideration of its part in a supportable and environment cognizant future.

Notwithstanding ecological worries, the general wellbeing ramifications of coal ignition are a basic part of the more extensive talk. Respiratory illnesses, like asthma and constant obstructive aspiratory sickness (COPD), are pervasive in networks presented to coal-terminated power plant outflows. Weak populaces, including kids and the older, face elevated chances, with concentrates reliably connecting coal-related air contamination to antagonistic wellbeing results.

Moreover, the ecological equity aspects of coal-related medical problems can't be disregarded. Frequently, coal-terminated power plants are arranged in nearness to underestimated networks, where occupants might come up short on assets to move or access sufficient medical services. This unbalanced weight on weak populaces highlights the requirement for evenhanded natural arrangements that focus on the prosperity, everything being equal, paying little heed to financial status or segment factors.

The progress towards sustainable power sources arises as a significant answer for moderate the natural and wellbeing concerns related with coal. Embracing sun based, wind, hydropower, and other clean energy choices not just addresses the dire need to battle environmental change yet additionally advances a better and more maintainable future. In any case, this progress requires a purposeful exertion from policymakers, enterprises, and networks to explore the monetary, social, and political difficulties related with such a significant shift.

Taking everything into account, the natural and wellbeing concerns connected to coal mining and consuming are diverse and interconnected. From the biological interruptions made by mining exercises the air and water contamination coming about because of coal burning, the adverse consequences length neighborhood and worldwide scales. The wellbeing results, especially respiratory and cardiovascular infections, highlight the pressing need to progress towards cleaner and more practical energy sources. As the world wrestles with the goals of environmental change and natural equity, a reconsideration of our reliance on coal becomes basic for a future where energy creation lines up with biological versatility and human prosperity.

3.2 Introduction of alternative energy sources and the push for cleaner technologies.

The 21st century has introduced a significant arousing to the ecological difficulties presented by conventional energy sources, inciting a worldwide change in perspective toward option and cleaner innovations. As social orders wrestle with the outcomes of environmental change, air contamination, and limited petroleum derivative holds, the basic to embrace economical energy arrangements has become progressively clear. This shift addresses a reaction to the ecological emergencies as well as a potential chance to rethink the direction of human turn of events and address the pressing requirement for energy security.

Environmentally friendly power sources stand at the front of this change, offering a commitment of perfect, bountiful, and maintainable power. Sun oriented energy, outfit from the sun's beams, has arisen as an encouraging sign in the journey for cleaner innovations. Photovoltaic cells convert daylight into power, giving a sustainable and boundless wellspring of force. The progressions in sun based innovation have prompted expanded productivity and decreased costs, making sun oriented energy an undeniably practical choice for both enormous scope power plants and decentralized, off-network frameworks.

Essentially, wind energy has seen exceptional development as wind turbines bridle the active energy of moving air to create power. Wind power is a developed and quickly extending environmentally friendly power source that has demonstrated its viability in different geological locales. Seaward wind ranches, specifically, gain serious areas of strength for by predictable breezes over waterways, introducing an immense undiscovered capacity for clean energy creation. The adaptability of wind power — from little turbines for individual families to gigantic breeze ranches for framework coordination — highlights its flexibility as a central member in the progress toward maintainable energy.

Hydropower, got from the gravitational power of streaming water, has for quite some time been a foundation of sustainable power. From customary water wheels to current hydroelectric dams, the saddling of water's energy has fueled networks and enterprises. While enormous scope hydroelectric ventures ecologically affect stream environments and removal of networks, creative methodologies, for example, run-of-stream and limited scope hydroelectric frameworks, plan to limit these downsides and advance more economical utilization of water assets.

Geothermal energy, drawn from the World's inside heat, addresses one more feature of the spotless energy scene. Using heat from the World's outside layer for power age and direct warming applications, geothermal power plants offer a dependable and steady wellspring of energy. The improvement of upgraded geothermal frameworks (EGS) and other creative innovations further extends the capability of geothermal energy, making it a significant supporter of the enhanced energy portfolio required for a manageable future.

Bioenergy, got from natural materials like plants and waste, gives one more road to cleaner advances. Biomass, biofuels, and biogas offer energy options with lower carbon impressions contrasted with conventional petroleum products. Biomass energy, acquired from natural materials like wood, farming buildups, and devoted energy crops, can be combusted straightforwardly or changed over into biofuels for transportation. Biogas, created through the anaerobic processing of natural waste, presents a sustainable wellspring of methane for power and intensity age. The reasonable administration of bioenergy assets, in any case, requires cautious thought to forestall deforestation, soil debasement, and rivalry with food crops.

The turn of events and combination of these elective energy sources are catalyzed by progressions in energy capacity advancements. Energy capacity arrangements, for example, batteries and other arising advances, assume a vital part in tending to the irregular idea of sustainable power. Batteries, specifically, empower the capacity of overabundance energy created during top creation periods, which can be delivered during times of appeal or low environmentally friendly power accessibility. The advancement of energy stockpiling innovations not just upgrades the unwavering quality of sustainable power sources yet in addition adds to the formation of stronger and decentralized energy frameworks.

The progress to cleaner innovations reaches out past the domain of power age to incorporate the transportation area. Electric vehicles (EVs) have arisen as an extraordinary power, offering an economical option in contrast to conventional gas powered motor vehicles. The jolt of transportation decreases dependence on non-renewable energy sources, mitigates air contamination, and shortens ozone depleting substance outflows. The developing reception of electric vehicles, combined with headways in battery innovation and charging framework, denotes a crucial second in the journey for economical and cleaner transportation.

Couple with the ascent of sustainable power and electric vehicles, brilliant networks have arisen as a basic part of the cutting edge energy scene. Shrewd lattices influence cutting edge innovations, including sensors, correspondence organizations, and information investigation, to upgrade the age, dispersion, and utilization of power. By upgrading lattice strength, empowering continuous observing, and working with the joining of environmentally friendly power sources, brilliant networks make ready for a more effective and economical energy framework.

The push for cleaner advances isn't exclusively determined by ecological contemplations yet is progressively perceived as a financial objective. The environmentally friendly power area has turned into a strong motor for work creation, encouraging development, and driving monetary development. As the expenses of sustainable advancements keep on declining, the progress to cleaner energy sources is progressively viewed as a financially feasible and socially dependable decision. Interest in clean energy mitigates natural effect as well as positions countries and businesses at the cutting edge of an expanding worldwide market for feasible advancements.

Worldwide drives and arrangements further highlight the worldwide obligation to cleaner advancements. The Paris Understanding, embraced in 2015, addresses a milestone accord that joins countries in their endeavors to restrict an Earth-wide temperature boost and change to a low-carbon future. The understanding underscores the significance of environmentally friendly power, energy effectiveness, and supportable turn of events, flagging an aggregate assurance to address the interconnected difficulties of environmental change, energy access, and natural manageability.

While the force toward cleaner innovations is empowering, provokes continue on the way to a supportable energy future. The discontinuity of environmentally friendly power sources represents a one of a kind arrangement of difficulties for lattice steadiness and unwavering quality. To address this, continuous innovative work endeavors center around energy capacity, lattice enhancement, and the advancement of imaginative answers for balance organic market in an undeniably sustainable driven energy scene.

Moreover, the eliminating of existing non-renewable energy source foundation presents financial and social difficulties. Networks and businesses dependent on conventional energy sources might confront financial separation, requiring an equitable and comprehensive change that focuses on work retraining, monetary enhancement, and the making of new open doors in the perfect energy area.

The job of policymakers is fundamental in exploring this change. States overall should sanction strategies that boost the reception of cleaner advancements, deter dependence on petroleum products, and establish a climate helpful for economical energy improvement.

Administrative systems, sponsorships, and market components assume a critical part in molding the direction of the energy scene, impacting speculation choices and directing the improvement of cleaner advances.

All in all, the presentation of elective energy sources and the push for cleaner advances address a urgent point in mankind's set of experiences. The basic to address ecological difficulties, relieve environmental change, and guarantee energy security has catalyzed a worldwide shift toward manageable and sustainable power arrangements. From sun oriented and wind capacity to electric vehicles and brilliant matrices, the different cluster of cleaner innovations offers a pathway to a future where human improvement is fit with the natural flexibility of the planet. As countries, enterprises, and networks team up to embrace this extraordinary excursion, the vision of a cleaner, more maintainable energy future becomes a chance as well as a basic for the prosperity of current and people in the future.

3.3 Initial steps towards reducing coal dependence and promoting sustainable energy.

The basic to address the natural and wellbeing concerns related with coal mining and consuming has incited a worldwide shift towards diminishing coal reliance and advancing reasonable energy choices. This progress addresses a diverse test that traverses strategy drives, mechanical development, and cultural change. As countries wrestle with the critical need to moderate environmental change and embrace cleaner advances, the underlying strides towards lessening coal reliance are urgent in setting the direction for a manageable energy future.

Strategy mediations assume a focal part in guiding the change away from coal. Legislatures overall are perceiving the basic to institute and implement guidelines that check coal use, advance environmentally friendly power reception, and boost economical practices. A key strategy instrument is the execution of outflows norms and guidelines that limit how much toxins delivered into the climate during coal ignition.

These norms act as a vital component to decrease air contamination and moderate the wellbeing influences related with coal-terminated power plants.

All the while, policymakers are acquainting market instruments with incorporate the outside expenses of coal utilization. Carbon valuing, through components like carbon assessments or cap-and-exchange frameworks, gives monetary impetuses to ventures to decrease their fossil fuel byproducts. By putting a cost on carbon, states make a monetary disincentive for coal utilization, empowering a shift towards cleaner and more feasible energy sources.

The getting rid of coal appropriations is another significant approach measure. Diverting monetary help away from the coal business and towards sustainable power drives assists level the playing with handling, making clean advancements all the more financially cutthroat.

These strategy shifts convey clear messages to enterprises and financial backers about the public authority's obligation to a low-carbon future, cultivating the important circumstances for the progress away from coal.

Moreover, administrative systems that work with the coordination of environmentally friendly power into the network are fundamental. Strategies supporting the improvement of brilliant frameworks, energy capacity innovations, and network adaptability empower a smoother progress by tending to the discontinuous idea of sustainable sources. Coordinating renewables into the current energy framework requires an administrative climate that supports development and interests in network modernization.

On the global stage, cooperative endeavors are in progress to speed up the decrease of coal reliance. Arrangements, for example, the Driving Past Coal Collusion unite legislatures, organizations, and associations focused on progressively getting rid of unabated coal power and supporting a simply change for impacted networks. The union gives a stage to shared learning, best practices, and cooperative drives to drive the worldwide change away from coal.

Notwithstanding strategy measures, mechanical development assumes a significant part in lessening coal reliance. Headways in environmentally friendly power advances have made them more productive, practical, and versatile, adding to their expanded reception. Sun based photovoltaic (PV) and wind turbine advancements, specifically, have seen huge improvement, with upgrades in effectiveness and decreases in assembling costs.

Energy capacity innovations, like batteries, are basic for tending to the irregular idea of environmentally friendly power sources. Forward leaps in battery innovation improve the ability to store abundance energy created during top creation periods, making it accessible during seasons of low sustainable power accessibility. These advancements add to the unwavering quality and solidness of sustainable power frameworks, further working with the change away from coal.

Innovative work endeavors are additionally centered around improving the proficiency of existing inexhaustible advances and investigating arising wellsprings of clean energy. Developments in concentrated sun oriented power, cutting edge breeze turbines, and high level geothermal frameworks add to the enhancement of the spotless energy portfolio. As these advancements mature, they become progressively suitable choices to coal, offering manageable answers for meeting the world's developing energy interest.

Past power age, the transportation area addresses a huge field for lessening coal reliance. The charge of transportation through electric vehicles (EVs) has built up some decent momentum as a compelling methodology to lessen dependence on petroleum derivatives.

States and ventures are boosting the reception of EVs through sponsorships, tax breaks, and the development of charging foundation. The progress to electric versatility diminishes the interest for oil as well as lines up with more extensive endeavors to decarbonize the whole energy framework.

In the modern area, which frequently depends vigorously on coal for intensity and power, the arrangement of clean advancements is essential. Zap of modern cycles, combined with the utilization of environmentally friendly power sources, offers a pathway to lessen the carbon impression of modern exercises. Furthermore, the reception of energy-productive advancements and feasible practices assists businesses with limiting their dependence on coal and add to in general discharges decrease.

Local area commitment and mindfulness assume an essential part in pushing the progress away from coal. Instructing people group about the natural, wellbeing, and financial effects of coal mining and consuming cultivates an aggregate comprehension of the requirement for change. Local area drove drives, for example, environmentally friendly power cooperatives and decentralized energy projects, engage neighborhood populaces to effectively take part in the change and advantage monetarily from the shift towards reasonable energy.

Guaranteeing a simply progress for networks generally reliant upon coal is a basic part of these endeavors. As coal-subordinate districts face monetary disengagement, employment misfortunes, and local area disturbance, policymakers should execute procedures that focus on work retraining, financial enhancement, and the formation of new open doors in the spotless energy area. Social projects, monetary help, and local area commitment drives are fundamental parts of a simply change that limits the adverse consequences on impacted networks.

Monetary foundations and financial backers likewise assume a urgent part in pushing the change away from coal. Divestment from coal-related ventures and interests in clean energy portfolios convey a strong message about the monetary dangers related with petroleum derivative resources. The rising accessibility of green securities and maintainable venture reserves gives roads to financial backers to help projects lined up with ecologically capable practices and environmentally friendly power advancement.

Worldwide collaboration is basic in tending to the transboundary idea of natural difficulties related with coal. Cooperative drives, innovation move, and limit building programs work with the sharing of information and assets among countries. Created nations can offer help to emerging countries in their endeavors to progress to cleaner advances, encouraging a worldwide local area focused on a manageable and fair energy future.

All in all, the underlying strides towards decreasing coal reliance and advancing economical energy envelop an exhaustive methodology that traverses strategy, innovation, and cultural commitment.

State run administrations, businesses, and networks should work cooperatively to order and implement approaches that boost the change away from coal, while mechanical developments and interests in clean energy arrangements drive the practicality and adaptability of these other options. By tending to the ecological, wellbeing, and financial difficulties related with coal, the world can make ready for a maintainable energy future that focuses on the prosperity of the planet and its occupants.

The worldwide scene is seeing a seismic change in the energy worldview as countries wrestle with the basic to address the ecological and wellbeing concerns related with coal reliance. This change addresses a perplexing exchange of strategy drives, mechanical headways, and cultural changes, all pointed toward cultivating reasonable other options. As the unfriendly effects of coal mining and consuming become progressively obvious — from air contamination and respiratory infections to environmental change and living space annihilation — drives to lessen coal reliance and advance reasonable energy are building up forward movement across the globe.

Strategy Intercessions:

At the very front of this change are strategy intercessions intended to reshape the energy scene. States are perceiving the dire need to order rigid guidelines that limit outflows from coal-terminated power plants, tending to the air contamination that has long tormented networks in nearness to such offices. Outflow principles act as a pivotal device, directing the satisfactory degrees of toxins delivered into the climate during coal burning. These principles relieve air contamination as well as add to general wellbeing by diminishing the commonness of respiratory sicknesses related with coal outflows.

Carbon evaluating components, for example, carbon expenses and cap-and-exchange frameworks, address one more approach road to incorporate the outer expenses of coal use. By putting a cost on fossil fuel byproducts, legislatures make financial impetuses for enterprises to progress toward cleaner innovations. This market-driven approach energizes the decrease of ozone depleting substance emanations, encouraging a more feasible energy scene.

Deliberately eliminating endowments for the coal business is a fundamental arrangement measure in making everything fair for sustainable power sources. Monetary help diverted away from coal and towards cleaner choices helps make supportable advancements monetarily cutthroat. This redirection lines up with more extensive financial objectives and conveys a reasonable message to ventures and financial backers about the public authority's obligation to a low-carbon future.

Besides, administrative systems that work with the mix of environmentally friendly power into existing frameworks are fundamental. Savvy framework drives, which influence cutting edge innovations, for example, sensors and information investigation, upgrade the age, circulation, and utilization of power.

These systems upgrade lattice flexibility, empower continuous observing, and work with the coordination of environmentally friendly power sources, tending to the irregular idea of these perfect other options.

Globally, cooperative endeavors, for example, the Controlling Past Coal Collusion unite states, organizations, and associations focused on transitioning away from unabated coal power. This union not just sets a common obligation to decreasing coal reliance yet additionally gives a stage to information sharing, prescribed procedures, and cooperative drives on a worldwide scale.

Mechanical Advancement:

The decrease of coal reliance is characteristically connected to mechanical development, which assumes a critical part in progressing reasonable other options. The development of environmentally friendly power innovations, for example, sun oriented photovoltaic (PV) and wind turbines, has made them more proficient and practical. Enhancements in assembling cycles and materials have added to the expanded reception of these advances, both for an enormous scope for network mix and on a more limited size for decentralized, off-matrix arrangements.

Energy capacity innovations, especially batteries, address the discontinuity of sustainable power sources. Leap forwards in battery innovation improve the ability to store abundance energy created during top creation periods, making it accessible during times of low environmentally friendly power accessibility. These headways add to the dependability and steadiness of environmentally friendly power frameworks, guaranteeing a persistent and reliable power supply.

Innovative work endeavors are additionally centered around upgrading the productivity of existing sustainable advancements and investigating arising wellsprings of clean energy. Developments in concentrated sunlight based power, cutting edge breeze turbines, and high level geothermal frameworks add to the expansion of the spotless energy portfolio. As these advances mature, they become progressively practical choices to coal, offering supportable answers for meeting the world's developing energy interest.

In the transportation area, the charge of vehicles stands apart as an extraordinary mechanical arrangement. Electric vehicles (EVs) offer a maintainable option in contrast to customary gas powered motor vehicles, diminishing the dependence on petroleum products and controlling air contamination. Headways in battery innovation, combined with the extension of charging foundation, are key drivers in making EVs a pragmatic and far and wide answer for decrease coal reliance in the vehicle area.

In the modern area, which generally depends vigorously on coal for intensity and power, the organization of clean advancements is vital. Zap of modern cycles, combined with the utilization of sustainable power sources, offers a pathway to lessen the carbon impression of modern exercises. Also, the reception of energy-proficient advances and reasonable practices assists ventures with limiting their dependence on coal and add to generally emanations decrease.

Local area Commitment and Simply Change:

Local area commitment and mindfulness structure a crucial part of the change away from coal. Instructing people group about the ecological, wellbeing, and financial effects of coal mining and consuming encourages an aggregate comprehension of the requirement for change. Local area drove drives, for example, environmentally friendly power cooperatives and decentralized energy projects, engage neighborhood populaces to effectively partake in the progress and advantage financially from the shift towards practical energy.

Guaranteeing a simply change for networks generally reliant upon coal is a basic part of these endeavors. As coal-subordinate locales face financial disengagement, employment misfortunes, and local area disturbance, policymakers should carry out methodologies that focus on work retraining, monetary expansion, and the formation of new open doors in the spotless energy area. Social projects, monetary help, and local area commitment drives are fundamental parts of a simply change that limits the adverse consequences on impacted networks.

Monetary Movements and Worldwide Participation:

Monetary organizations and financial backers assume a urgent part in pushing the progress away from coal. Divestment from coal-related tasks and interests in clean energy portfolios convey a strong message about the monetary dangers related with petroleum product resources. The rising accessibility of green securities and reasonable venture reserves gives roads to financial backers to help projects lined up with earth capable practices and environmentally friendly power advancement.

Worldwide participation is basic in tending to the transboundary idea of ecological difficulties related with coal. Cooperative drives, innovation move, and limit building programs work with the sharing of information and assets among countries. Created nations can offer help to emerging countries in their endeavors to change to cleaner innovations, encouraging a worldwide local area focused on a maintainable and impartial energy future.

Difficulties and Future Standpoint:

While the underlying strides towards diminishing coal reliance and advancing economical energy are promising, moves endure on the way to a completely manageable energy future. The discontinuity of environmentally friendly power sources represents a special arrangement of difficulties for lattice soundness and unwavering quality.

Progressing innovative work endeavors center around energy capacity, lattice enhancement, and the improvement of imaginative answers for balance organic market in an undeniably sustainable driven energy scene.

The eliminating of existing petroleum product foundation presents monetary and social difficulties. Networks and businesses dependent on conventional energy sources might confront financial separation, requiring a fair and comprehensive progress that focuses on work retraining, monetary expansion, and the making of new open doors in the perfect energy area.

The job of policymakers stays foremost in exploring this progress. State run administrations overall should keep on establishing arrangements that boost the reception of cleaner innovations, deter dependence on petroleum products, and establish a climate helpful for economical energy advancement. Administrative systems, sponsorships, and market components assume an essential part in forming the direction of the energy scene, impacting speculation choices and directing the improvement of cleaner innovations.

Chapter 4

Technologies in Transition

In the consistently developing scene of innovation, we end up in the midst of a significant progress — a period set apart by phenomenal headways and groundbreaking movements. This unique transaction among development and variation has led to a large number of innovations that are forming the manner in which we live, work, and interface with the world.

At the front of this change is the coming of man-made consciousness (computer based intelligence), a field that has seen striking steps as of late. Simulated intelligence, once bound to the domain of sci-fi, has turned into an essential piece of our regular routines. AI calculations, fueled by huge datasets, empower PCs to learn and settle on choices without express programming. This capacity has saturated different ventures, from medical care to fund, upsetting how assignments are achieved and experiences are determined.

Couple with simulated intelligence, the Web of Things (IoT) has arisen as a vital participant in the mechanical scene. IoT flawlessly interconnects gadgets, from home devices to modern hardware, making a trap of information driven correspondence. This interconnectivity works with the assortment and investigation of continuous information, prompting further developed proficiency, improved independent direction, and the advancement of shrewd biological systems. The combination of simulated intelligence and IoT has led to clever frameworks that can independently answer evolving conditions, preparing for an additional associated and robotized future.

Blockchain innovation, initially considered as the fundamental framework for digital currencies, has risen above its underlying reason. Past the domain of money, blockchain is being saddled for its capability to make straightforward, secure, and decentralized frameworks. Its decentralized nature makes it impervious to altering, guaranteeing the respectability of information across different applications. From store network the executives to advanced personality confirmation, blockchain is cultivating trust and straightforwardness in assorted areas.

The energy scene is going through a significant change with the rising reception of sustainable power sources. Sun based and wind power, when thought about elective choices, are presently essential parts of the worldwide energy blend. Propels in energy capacity advancements are tending to the irregular idea of sustainable sources, empowering the consistent mix of clean energy into existing matrices. The progress towards environmentally friendly power isn't just determined by ecological contemplations yet additionally by the financial reasonability of maintainable other options.

Biotechnology is another outskirts that is encountering quick change. CRISPR-Cas9, a progressive quality altering device, has opened additional opportunities in the field of hereditary qualities. The capacity to exactly alter qualities holds the possibility to kill hereditary sicknesses, improve farming efficiency, and even rethink the conceivable outcomes of human advancement. Moral contemplations encompassing the utilization of such useful assets highlight the requirement for capable and straightforward practices in the domain of biotechnology.

As innovation reshapes enterprises, the idea of work is likewise going through a change in perspective. The ascent of robotization, filled by advanced mechanics and man-made intelligence, is modifying the conventional scene of work. While robotization brings expanded proficiency and efficiency, it likewise raises worries about work dislodging and the requirement for reskilling the labor force. Finding some kind of harmony between mechanical development and the human labor force is a basic test that requires smart thought.

The advanced insurgency has saturated the domains of money and business, leading to fintech and online business biological systems. Versatile installment arrangements, blockchain-based exchanges, and advanced monetary forms are reclassifying the manner in which we manage monetary exchanges.

Internet business stages have changed the retail scene, offering shoppers uncommon comfort and admittance to a worldwide commercial center. The combination of innovation and money is reshaping customary financial models and making ready for a more comprehensive and productive monetary biological system.

Online protection has turned into an undeniably basic part of the innovative scene. As our dependence on advanced foundation develops, so does the significance of protecting it from digital dangers. The multiplication of associated gadgets and the tremendous measures of information produced make new weaknesses that require vigorous network protection measures. From cutting edge encryption advances to man-made intelligence driven danger recognition, the network safety scene is developing to neutralize the always changing strategies of vindictive entertainers.

The field of medical care is seeing an upset driven by innovation. Telemedicine, fueled by fast web and high level specialized devices, has made medical services more open to remote and underserved populaces. The mix of simulated intelligence in diagnostics and therapy arranging is improving the precision and effectiveness of clinical practices. Wearable gadgets and wellbeing following applications engage people to play a functioning job in dealing with their wellbeing. The convergence of innovation and medical services holds the commitment of customized medication and worked on tolerant results.

Progressions in space investigation are pushing the limits of human information and capacity. Privately owned businesses, notwithstanding conventional space organizations, are effectively adding to the investigation of space. From reusable rocket innovation to plans for laying out human settlements on different planets, the space business is encountering a renaissance. The mission to grasp the universe and investigate divine bodies mirrors humankind's inborn interest and assurance to stretch the boundaries of what is conceivable.

Instruction is going through a change worked with by computerized innovations. Internet learning stages, virtual study halls, and instructive applications are reshaping how information is obtained and shared. The democratization of schooling is turning into a reality as innovation separates obstructions to get to. The idea of deep rooted learning is acquiring conspicuousness, underlining the requirement for constant expertise advancement in a period of fast mechanical change.

The crossing point of innovation and supportability is a critical topic in the continuous progress. Feasible advancements, going from sustainable power answers for eco-accommodating materials, are getting momentum as the worldwide local area wrestles with natural difficulties. Round economy models, which focus on asset proficiency and waste decrease, are being embraced by organizations and legislatures the same. The quest for maintainability through innovation mirrors an acknowledgment of the interconnectedness between human exercises and the strength of the planet.

As we explore this time of innovative progress, moral contemplations pose a potential threat. The mindful turn of events and organization of advances require a cautious harmony among development and moral contemplations. Questions encompassing information protection, algorithmic inclination, and the moral ramifications of strong advances highlight the requirement for moral structures and administrative measures. The mindful utilization of innovation is fundamental to guarantee that the advantages of advancement are impartially conveyed and that potentially negative side-effects are relieved.

The speed of mechanical headway makes it clear that things are not pulling back. Quantum registering, with its capability to reform computational power, is not too far off. 5G innovation is ready to introduce another period of availability, empowering quicker and more dependable correspondence. Expanded reality (AR) and computer generated reality (VR) advances are changing the manner in which we experience and cooperate with the world. These arising advances hold the commitment of reshaping enterprises and human encounters in manners that were once unfathomable.

All in all, the continuous mechanical progress is a demonstration of human resourcefulness and our ability to develop. The intermingling of artificial intelligence, IoT, blockchain, environmentally friendly power, biotechnology, and other groundbreaking advancements is reshaping the structure holding the system together. As we outfit the capability of innovation, moving toward its turn of events and sending with a feeling of obligation and moral consideration is basic. Our decisions today will shape the direction of our aggregate future, and by exploring this progress insightfully, we can open a future where innovation fills in as a power for positive change and human thriving.

4.1 Overview of emerging technologies for cleaner coal extraction and combustion.

The journey for cleaner and more economical energy sources has prodded continuous innovative work in the field of coal extraction and ignition. Regardless of the rising spotlight on sustainable power, coal stays a huge piece of the worldwide energy blend, especially in locales vigorously subject to this petroleum product. In light of natural worries and the need to moderate the effect of coal-based energy creation, a scope of arising advances is being investigated to upgrade the proficiency of coal extraction and burning while at the same time limiting ecological outcomes.

One outstanding area of development spins around cutting edge coal extraction innovations. Conventional coal mining techniques have frequently brought about ecological corruption, including territory disturbance, water contamination, and the arrival of methane — a powerful ozone harming substance. To resolve these issues, specialists and industry partners are investigating advances that can separate coal all the more effectively and with diminished natural effect.

In-situ gasification is one such innovation that holds guarantee for cleaner coal extraction. Dissimilar to customary mining strategies, in-situ gasification includes changing over coal underground into a burnable gas. This cycle might possibly lessen the natural impression related with surface mining while at the same time expanding energy recuperation. Nonetheless, difficulties, for example, overseeing subsurface responses and guaranteeing ecological wellbeing should be painstakingly tended to for boundless reception.

Carbon catch and capacity (CCS) is one more basic part of the change towards cleaner coal. CCS innovations mean to catch the carbon dioxide (CO_2) emanations delivered during coal burning and store them underground, forestalling their delivery into the environment. The fruitful execution of CCS can altogether decrease the natural effect of coal-terminated power plants and broaden the life expectancy of existing foundation.

Notwithstanding CCS, progressions in carbon catch and use (CCU) are getting forward momentum. CCU includes catching CO_2 discharges and changing over them into significant items, like synthetic substances, fills, or development materials.

This approach not just mitigates the natural effect of CO2 emanations yet in addition changes them into assets, adding to a more round and economical economy.

Endeavors are likewise in progress to upgrade the effectiveness of coal ignition itself. High-effectiveness, low-discharge (HELE) advances expect to enhance the ignition cycle, bringing about expanded energy yield and diminished outflows. Supercritical and super supercritical steam cycles, for instance, work at higher temperatures and tensions, working on the proficiency of coal-terminated power plants. These innovations not just add to a more economical utilization of coal yet additionally line up with worldwide endeavors to lessen ozone depleting substance discharges.

Fluidized bed ignition is one more creative way to deal with coal burning that offers benefits as far as proficiency and outflows control. In fluidized bed ignition, coal particles are suspended in a bed of idle material, making a dynamic and productive burning cycle. This innovation empowers better control of ignition conditions, prompting lower outflows of sulfur dioxide (SO2) and nitrogen oxides (NOx). Also, fluidized bed burning takes into account the immediate catch of poisons, adding to cleaner coal ignition.

The mix of sustainable power sources with coal-terminated power plants is one more road for improving the supportability of coal-based energy age. Half and half frameworks that consolidate coal and sustainable power, for example, sun based or wind, offer a more adjusted and harmless to the ecosystem approach. By utilizing the qualities of both regular and environmentally friendly power sources, these crossover frameworks can give a persistent and solid power supply while diminishing by and large ecological effect.

The idea of a coal-terminated power plant with incorporated biomass transformation is acquiring consideration too. Biomass, for example, horticultural buildups or devoted energy crops, can be co-terminated with coal to deliver power. This approach not just diminishes the carbon impression of coal-terminated power age yet additionally upholds the improvement of a bioenergy area. The co-terminating of biomass and coal addresses a temporary system that can add to a more reasonable energy blend.

Progressions in material science are assuming a pivotal part in working on the proficiency and ecological execution of coal-based advancements. Superior execution materials that can endure the brutal states of coal ignition, like high temperatures and destructive conditions, are fundamental for the improvement of cutting edge coal-terminated power plants. Research in materials science means to upgrade the solidness and dependability of parts, at last broadening the functional existence of coal-based offices.

Moreover, the utilization of computerized reasoning (simulated intelligence) and high level control frameworks is turning out to be progressively pervasive in the streamlining of coal burning cycles. Computer based intelligence calculations can break down immense measures of information continuously, taking into account dynamic changes in accordance with ignition conditions. This degree of accuracy further develops productivity as well as empowers more powerful discharges control. The use of man-made intelligence in coal-terminated power plants lines up with the more extensive pattern of digitalization in the energy area, cultivating a more canny and versatile way to deal with power age.

The worldwide shift towards a hydrogen-based economy is impacting the coal area too. Hydrogen delivered from coal, known as blue hydrogen when joined with carbon catch, can possibly act as a cleaner option in contrast to conventional coal burning. The course of hydrogen creation from coal includes gasification, trailed by the division of hydrogen and carbon dioxide. The caught CO_2 can then be put away or used, adding to the generally ecological supportability of the coal-to-hydrogen pathway.

In the domain of coal-to-fluids (CTL) innovation, progressions are centered around delivering fluid fills from coal in an earth capable way. CTL processes include coal gasification followed by the combination of fluid hydrocarbons. Specialists are investigating strategies to work on the productivity of these cycles and lessen the ecological effect, especially regarding water utilization and ozone harming substance discharges.

The progress to cleaner coal extraction and burning isn't exclusively a mechanical test; it likewise includes social, financial, and strategy contemplations. The acknowledgment and reception of these arising innovations rely upon a mix of specialized practicality, financial reasonability, and administrative systems that boost feasible practices. Strategy mediations, for example, carbon estimating and support for innovative work, assume a crucial part in molding the direction of the coal area towards cleaner and more maintainable practices.

Worldwide cooperation and information sharing are fundamental parts of tending to the worldwide difficulties related with coal extraction and burning. Research foundations, industry partners, and policymakers from around the world are cooperating to trade experiences, best practices, and advancements. This cooperative methodology cultivates an aggregate comprehension of the intricacies engaged with progressing the coal area towards more noteworthy manageability.

All in all, the outline of arising innovations for cleaner coal extraction and burning mirrors a complex and dynamic scene. From cutting edge extraction strategies to high-proficiency ignition innovations, the coal area is going through a change pointed toward accommodating energy needs with natural worries. The reconciliation of environmentally friendly power, the improvement of elite execution materials, and the utilization of computerized innovations all add to a more maintainable and mindful utilization of coal. As the worldwide local area looks for a harmony between energy security and ecological stewardship, the continuous endeavors to improve inside the coal area highlight the significance of a thorough and cooperative way to deal with molding the eventual fate of energy.

4.2 Introduction to carbon capture and storage (CCS) technologies.

The raising worldwide worry over environmental change has strengthened endeavors to create and carry out procedures for relieving carbon dioxide (CO2) emanations. Among the variety of innovations pointed toward decreasing ozone harming substance discharges, Carbon Catch and Capacity (CCS) has arisen as a vital device in the journey for a feasible and low-carbon future. CCS advances are intended to catch CO2 outflows from modern cycles and power age, forestall their delivery into the environment, and safely store the caught CO2 underground. This extensive methodology tends to the test of decreasing outflows from areas that are innately hard to decarbonize.

The center idea of CCS spins around catching CO2 at its source, commonly huge point sources like power plants and modern offices, before it is delivered very high. The caught CO2 is then shipped to a capacity site and infused into land developments, like drained oil and gas repositories or profound saline springs. The capacity interaction guarantees that the CO2 remains safely caught underground, forestalling its delivery and adding to the general decrease of climatic CO2 fixations.

The first move toward quite a while cycle is carbon catch, where CO2 is isolated from different gases created during burning or modern cycles. Post-burning catch, pre-ignition catch, and oxyfuel burning are the three fundamental ways to deal with carbon catch. Post-burning catch includes eliminating CO2 from the vent gas after ignition, regularly utilizing solvents or sorbents. Pre-ignition catch, then again, involves catching CO2 before burning by changing over petroleum derivatives into syngas and afterward isolating the CO2. Oxyfuel ignition includes consuming petroleum products in a climate of unadulterated oxygen, bringing about a vent gas stream overwhelmingly made out of CO2.

When caught, the CO2 should be shipped to a reasonable stockpiling site. Transportation strategies incorporate pipelines, ships, and trucks. Pipelines are a typical and productive method of transport for enormous amounts of CO2, particularly when the source and capacity destinations are in closeness. Ships and trucks are elective transportation choices, especially for remote or more limited size projects. The determination of transportation mode relies upon variables like distance, volume, and framework accessibility.

The capacity period of CCS includes infusing the caught CO2 into topographical arrangements profound underground. Drained oil and gas repositories, where oil or gas has been separated, offer normal extra rooms for CO2. Profound saline springs, land developments loaded up with pungent water, are additionally reasonable capacity destinations. Also, upgraded oil recuperation (EOR) includes infusing CO2 into drained oil repositories to separate extra oil while at the same time putting away CO2. The capacity cycle depends on the physical and synthetic properties of the land arrangements to guarantee the protected and super durable control of CO2.

CCS advances stand out enough to be noticed for their capability to diminish CO2 emanations from different areas altogether. The power age area, especially coal-terminated power plants, has been a point of convergence for CCS execution. Given the worldwide predominance of coal as an energy source, CCS in coal-terminated power plants addresses a basic technique for accomplishing outflows decrease targets. Past power age, CCS has applications in enterprises like concrete, steel, and synthetic assembling, where discharges are trying to dispense with through customary means.

The arrangement of CCS innovations isn't without difficulties and contemplations. One huge test is the monetary feasibility of CCS projects. The significant expenses related with the catch, transportation, and capacity of CO2 can make CCS monetarily troublesome without suitable impetuses or administrative systems. Legislatures and policymakers assume an essential part in offering the important help, for example, carbon valuing systems or endowments, to make CCS financially doable for businesses.

Public insight and acknowledgment additionally factor into the fruitful sending of CCS advancements. Concerns connected with the security and perpetual quality of CO2 stockpiling, as well as possible natural effects, should be addressed through straightforward correspondence and adherence to rigid wellbeing norms. Local area commitment and mindfulness drives are crucial for construct trust and understanding with respect to the advantages and dangers of CCS projects.

The administrative scene assumes a significant part in molding the sending of CCS advances. Clear and predictable administrative systems are important to give a steady climate to venture and execution.

The foundation of administrative principles for CO2 stockpiling, observing, and confirmation is basic to guaranteeing the viability and wellbeing of CCS projects. Also, worldwide coordinated effort is urgent, as CO2 discharges and their effects are worldwide difficulties that rise above public boundaries.

Headways in CCS advancements keep on being a focal point of innovative work. The quest for more proficient and savvy catch strategies, novel capacity procedures, and advancements in transportation are continuous. Research is likewise investigating the capability of direct air catch, a cycle that includes catching CO2 straightforwardly from the air. While direct air catch is in the beginning phases of improvement and countenances difficulties connected with energy utilization and cost, it holds guarantee as a corresponding way to deal with customary CCS.

The job of CCS in accomplishing environment objectives has been perceived at the worldwide level. The Paris Understanding, a milestone worldwide accord on environment activity, recognizes the significance of accomplishing a harmony between anthropogenic discharges and evacuations of ozone depleting substances. CCS is recognized as one of the advances that can add to accomplishing this equilibrium. As nations pursue their emanations decrease targets, CCS is situated as a vital part of the arrangement of arrangements expected to address the worldwide environment challenge.

All in all, carbon catch and capacity advancements address a pivotal pathway towards accomplishing critical decreases in CO2 outflows from modern cycles and power age. The exhaustive methodology of catching, shipping, and safely putting away CO2 tends to the inborn difficulties related with discharges from areas that are hard to decarbonize. While confronting monetary, administrative, and public insight challenges, CCS innovations offer a substantial and versatile answer for moderate the effect of anthropogenic CO2 emanations on the environment. As innovative work endeavors proceed, and as steady strategy systems are laid out, CCS is ready to assume an undeniably essential part in the worldwide endeavors to battle environmental change and progress to a practical and low-carbon future.

4.3 Highlight innovations in renewable energy sources and advancements in electricity generation.

In the continuous quest for a manageable and low-carbon energy future, developments in environmentally friendly power sources and headways in power age have become basic central focuses. The need to progress away from customary petroleum product based power age techniques to alleviate environmental change has prodded innovative work endeavors around the world. Sustainable power advancements tackle normal assets like daylight, wind, water, and geothermal intensity to produce power in manners that are harmless to the ecosystem and economical. The accompanying areas dig into a portion of the vital developments and headways in the domain of environmentally friendly power and power age.

Sun based Photovoltaics (PV):

Quite possibly of the most extraordinary advancement in sustainable power is the fast development of sun oriented photovoltaic (PV) innovation. Sun powered chargers, contained interconnected sun based cells, convert daylight straightforwardly into power. Throughout the long term, progressions in materials, producing cycles, and proficiency have fundamentally worked on the presentation of sun powered PV frameworks. The advancement of cutting edge sun based cells, including slender film innovations and perovskite sun oriented cells, holds guarantee for additional rising effectiveness and decreasing the expense of sun powered power.

Coordinated sunlight based advances have likewise acquired unmistakable quality, for example, building-incorporated photovoltaics (BIPV) and sun powered windows. BIPV consistently integrates sun powered components into building materials, for example, sun based rooftop tiles or sunlight based exteriors, improving energy age without compromising style. Sun powered windows, implanted with straightforward sun based cells, empower the gathering of daylight without deterring sees, making them appropriate for both private and business applications.

Concentrated Sunlight based Power (CSP):

Concentrated Sunlight based Power (CSP) is one more inventive way to deal with bridling sun oriented energy. CSP frameworks use mirrors or focal points to focus daylight onto a little region, regularly a recipient, which then, at that point, changes over the sun powered energy into heat. This nuclear power is utilized to create steam that drives turbines, producing power. CSP offers the upside of giving dispatchable power, as nuclear power can be put away for sometime in the future, tending to the discontinuity related with conventional sun based PV frameworks.

Progressions in CSP innovations incorporate the utilization of cutting edge materials for heat capacity, like liquid salt or earthenware particles. These materials consider higher temperature activity, expanding the proficiency of the transformation cycle. Novel CSP plans, for example, tower and allegorical box setups, keep on being refined to upgrade execution and diminish costs. The combination of CSP with nuclear power stockpiling frameworks adds to the improvement of solid and dispatchable sustainable power arrangements.

Wind Energy:

In the domain of wind energy, advancements are driving the improvement of bigger and more proficient breeze turbines. The pattern toward bigger turbines with higher center levels and longer sharp edges has prompted expanded energy catch and worked on cost-viability. Seaward wind ranches, arranged in vast waters, benefit from more grounded and more predictable breeze speeds, making them a critical supporter of the development of wind energy limit.

Drifting breeze turbines address a remarkable progression in seaward wind innovation. Not at all like conventional fixed-base designs, drifting turbines are fastened to the seabed, permitting organization in more profound waters. This development opens up immense fields of the sea for potential breeze energy projects, opening new open doors for saddling seaward wind assets.

Past customary level hub wind turbines, vertical-pivot wind turbines (VAWTs) are acquiring consideration for their novel plan and expected benefits in unambiguous applications. VAWTs can catch wind from any course without the requirement for arrangement, making them appropriate for metropolitan and dispersed energy settings. Innovative work in VAWT innovation expect to advance effectiveness and address difficulties related with versatility.

Hydropower and Marine Energy:

Hydropower, a deeply grounded environmentally friendly power source, keeps on seeing headways in turbine plan and proficiency. Cutting edge hydropower advances, incorporating low-head and in-stream turbines, mean to grow the scope of reasonable locales for hydropower projects. Siphoned capacity hydropower, which includes putting away energy by siphoning water uphill during times of low interest and delivering it to create power during top interest, stays an essential part of network soundness.

Marine energy, incorporating wave and flowing energy, is an arising outskirts with huge potential. Wave energy converters catch the energy from sea waves, while flowing energy frameworks bridle the dynamic energy from the back and forth movement of tides. Advancements in materials, gadget plan, and sending strategies are tending to the specialized difficulties related with marine energy and working on the unwavering quality and productivity of these frameworks.

Geothermal Energy:

Geothermal energy, got from the World's inner intensity, is a dependable and steady wellspring of force. Progressions in geothermal advances incorporate upgraded geothermal frameworks (EGS), which include the infusion of water into hot stone arrangements to make cracks and animate the progression of intensity. This development grows the topographical reach of geothermal power age past areas with normally happening high-temperature supplies.

Geothermal parallel cycle power plants, using low-temperature geothermal assets, are acquiring consideration for their productivity and ecological advantages. These frameworks utilize an optional liquid with a lower limit than water to catch heat and produce power. Research is progressing to work on the productivity of paired cycle power plants and investigate new techniques for separating heat from geothermal repositories.

Energy Capacity Advancements:

Headways in energy capacity advances are fundamental to tending to the irregular idea of environmentally friendly power sources and improving matrix unwavering quality. Lithium-particle batteries, broadly utilized for electric vehicles, are likewise central participants in fixed energy capacity applications. Continuous examination centers around further developing the energy thickness, cycle life, and cost-adequacy of lithium-particle batteries.

Past lithium-particle, a different exhibit of energy stockpiling innovations is being investigated. Stream batteries, which store energy in fluid electrolytes, offer versatility and long cycle life. Strong state batteries, with strong electrolytes supplanting conventional fluid electrolytes, show guarantee for higher energy thickness and wellbeing. Gravity-based energy capacity frameworks, for example, siphoned hydro and creative ideas like gravitational energy stockpiling, add to the advancement of huge scope and network associated capacity arrangements.

Brilliant Frameworks and Lattice Joining:

The improvement of brilliant frameworks addresses a change in perspective in power age and dissemination. Brilliant frameworks influence progressed correspondence and control advancements to streamline the mix of sustainable power sources, improve network flexibility, and empower request reaction. Decentralized energy frameworks, for example, microgrids, engage neighborhood networks to create, store, and deal with their energy, cultivating energy freedom and flexibility.

Lattice mix advances assume a vital part in dealing with the fluctuation of environmentally friendly power sources. High level lattice the board frameworks utilize prescient examination and ongoing checking to adjust organic market, guaranteeing matrix steadiness. Request side administration methodologies, including brilliant meters and season of-purpose valuing, engage customers to arrive at informed conclusions about their energy utilization, adding to in general framework effectiveness.

Man-made consciousness (computer based intelligence) and Digitalization:

The incorporation of man-made consciousness (computer based intelligence) and digitalization is reforming the activity and streamlining of power age and dissemination frameworks. Computer based intelligence calculations dissect immense measures of information, empowering prescient support, issue location, and ongoing streamlining of environmentally friendly power resources. Advanced twins, virtual reproductions of actual resources, work with displaying and reenactment, supporting the plan and activity of environmentally friendly power projects.

AI applications in sustainable power reach out to energy anticipating, where calculations foresee sustainable power age in view of weather patterns and different factors. This capacity upgrades matrix the executives and arranging, empowering utilities to adjust the joining of discontinuous inexhaustible sources with conventional power age.

Jolt of Transportation:

The jolt of transportation, especially the broad reception of electric vehicles (EVs), is reshaping the power scene. Electric vehicles serve for of maintainable transportation as well as disseminated energy capacity gadgets. Vehicle-to-framework (V2G) advances empower bidirectional energy stream among EVs and the network, permitting EVs to act as versatile energy assets during top interest periods.

Charging foundation advancements, including quick charging innovations and remote charging, are addressing obstructions to EV reception. Quick accusing stations of higher power yield lessen charging times, upgrading the accommodation of EVs. Remote charging advancements wipe out the requirement for actual connectors, smoothing out the charging system and expanding openness.

Mixture and Coordinated Energy Frameworks:

The mix of different sustainable power sources into half and half energy frameworks is getting some forward momentum. Half and half frameworks join reciprocal energy sources, for example, sun based and wind, to supply give a more steady and dependable power. Moreover, the coupling of environmentally friendly power with energy capacity and reinforcement generators makes strong and self-supporting microgrid frameworks.

Consolidated environmentally friendly power and capacity projects, frequently alluded to as incorporated energy frameworks, enhance the collaboration between various advances. Coordinating sun based or wind with energy capacity advances considers the persistent conveyance of power, conquering the fluctuation inborn in individual sustainable sources. These incorporated frameworks add to lattice soundness and backing the progress to a more decentralized and supportable energy foundation.

Headways in power age have been instrumental in reshaping the worldwide energy scene, cultivating manageability, proficiency, and ecological stewardship. As the interest for power keeps on rising, driven by populace development, urbanization, and the digitalization of social orders, creative innovations are assuming a urgent part in satisfying this need while limiting the natural effect. This complete outline investigates key headways in power age across different spaces.

1. **Sun based Photovoltaics (PV):** Sun oriented photovoltaic (PV) innovation has gone through amazing headways, changing it into one of the most unmistakable wellsprings of environmentally friendly power. Enhancements in sun based cell productivity, fabricating cycles, and material advances have fundamentally brought down the expense of sun oriented power. Arising advances, for example, perovskite sun based cells and couple sun powered cells, mean to drive proficiency limits further.

   Dainty film sun oriented modules, which utilize more slender semiconductor layers than conventional silicon-based modules, offer adaptability and simplicity of incorporation into different applications. Building-coordinated photovoltaics (BIPV) flawlessly integrate sun based components into engineering structures, empowering the age of power while keeping up with stylish allure. These advancements in sun powered PV innovation add to the far and wide reception of sun oriented energy and its combination into different areas, including private, business, and utility-scale projects.

2. **Wind Energy:** Advances in wind energy innovation have moved the development of coastal and seaward wind power. Bigger and more productive breeze turbines, portrayed by expanded center point levels and longer rotor edges, catch higher breeze speeds and create greater power. Seaward wind ranches, arranged in vast waters, benefit from more grounded and more predictable breeze assets, making them a vital supporter of the extension of wind energy limit.

   Drifting breeze turbines, fastened to the seabed in more profound waters, address an extraordinary headway in seaward wind innovation. This development opens up immense fields of the sea for potential breeze energy projects, tending to imperatives related with conventional fixed-base designs. Innovative work endeavors keep on refining wind turbine plans, enhance execution, and lessen costs, further setting wind energy as a serious and versatile environmentally friendly power source.

3. **Hydropower and Marine Energy:** Hydropower, a deeply grounded sustainable power source, has seen headways in turbine plan and productivity. Creative low-head and in-stream turbines plan to grow the scope of appropriate destinations for hydropower projects, permitting outfitting energy from waterways and streams with lower head levels. Siphoned capacity hydropower, a type of lattice scale energy capacity, assumes a critical part in adjusting power organic market. Marine energy, enveloping wave and flowing energy, addresses an arising wilderness with critical potential. Propels in materials, gadget plan, and arrangement

techniques are tending to specialized difficulties and working on the dependability and proficiency of these frameworks. Wave energy converters and flowing energy frameworks tackle the force of sea flows, adding to the broadening of environmentally friendly power sources and supporting the change to a more manageable energy blend.

4. **Geothermal Energy:** Geothermal energy, got from the World's interior intensity, has seen progressions in upgraded geothermal frameworks (EGS). EGS includes infusing water into hot stone developments to make cracks and animate the progression of intensity, growing the geological reach of geothermal power age past locales with normally happening high-temperature supplies. Geothermal double cycle power plants, using low-temperature geothermal assets, show proficiency enhancements and add to the usage of geothermal energy in different settings.

5. **Concentrated Sun based Power (CSP):** Concentrated Sun oriented Power (CSP) innovations influence mirrors or focal points to concentrate daylight onto a little region, creating heat that is utilized to deliver steam and drive turbines. Progressions in CSP incorporate the utilization of cutting edge materials for heat capacity, like liquid salt or artistic particles. These materials empower higher-temperature activity, working on the proficiency of the change interaction. Imaginative CSP plans, for example, tower and illustrative box setups, keep on being refined to upgrade execution and decrease costs.

6. **Energy Capacity Innovations:** Energy stockpiling innovations assume a basic part in tending to the irregularity of sustainable power sources and improving framework unwavering quality. Lithium-particle batteries, generally utilized for electric vehicles, are central members in fixed energy capacity applications. Continuous exploration centers around further developing the energy thickness, cycle life, and cost-adequacy of lithium-particle batteries.

   Past lithium-particle, a different cluster of energy stockpiling innovations is being investigated. Stream batteries, utilizing fluid electrolytes, offer adaptability and long cycle life. Strong state batteries, with strong electrolytes supplanting customary fluid electrolytes, show guarantee for higher energy thickness and wellbeing. Gravity-based energy capacity frameworks, including siphoned hydro and gravitational energy stockpiling, add to the improvement of huge scope and matrix associated capacity arrangements.

7. **Savvy Matrices and Lattice Coordination:** The improvement of brilliant networks denotes a change in perspective in power age and conveyance. Savvy matrices influence progressed correspondence and control advancements to upgrade the incorporation of sustainable power sources, improve framework versatility, and empower request reaction. Decentralized energy frameworks, for example, microgrids, engage neighborhood networks to create, store, and deal with their energy, cultivating energy autonomy and versatility.

   Framework mix advances assume a vital part in dealing with the changeability of sustainable power sources. High level framework the executives frameworks utilize prescient investigation and ongoing checking to adjust organic market, guaranteeing matrix dependability.

Request side administration techniques, including savvy meters and season of-purpose valuing, engage customers to settle on informed conclusions about their energy utilization, adding to generally speaking lattice proficiency.

8. **Man-made reasoning (artificial intelligence) and Digitalization:** The coordination of man-made consciousness (simulated intelligence) and digitalization is upsetting the activity and advancement of power age and circulation frameworks. Artificial intelligence calculations investigate tremendous measures of information, empowering prescient support, shortcoming discovery, and continuous advancement of sustainable power resources. Advanced twins, virtual imitations of actual resources, work with displaying and reproduction, supporting the plan and activity of sustainable power projects.

   AI applications in sustainable power stretch out to energy guaging, where calculations anticipate sustainable power age in view of atmospheric conditions and different factors. This capacity upgrades matrix the executives and arranging, empowering utilities to adjust the reconciliation of irregular inexhaustible sources with customary power age.

9. **Zap of Transportation:** The zap of transportation, especially the far and wide reception of electric vehicles (EVs), is reshaping the power scene. Electric vehicles serve for of supportable transportation as well as dispersed energy capacity gadgets. Vehicle-to-framework (V2G) advances empower bidirectional energy stream among EVs and the lattice, permitting EVs to act as portable energy assets during top interest periods.

   Charging framework advancements, including quick charging innovations and remote charging, are addressing boundaries to EV reception. Quick accusing stations of higher power yield diminish charging times, upgrading the comfort of EVs. Remote charging advances dispose of the requirement for actual connectors, smoothing out the charging system and expanding availability.

10. **Half and half and Incorporated Energy Frameworks:** The combination of different sustainable power sources into mixture energy frameworks is building up forward momentum. Half and half frameworks consolidate corresponding energy sources, for example, sun based and wind, to supply give a more steady and solid power. Furthermore, the coupling of environmentally friendly power with energy capacity and reinforcement generators makes versatile and self-supporting microgrid frameworks.

Joined environmentally friendly power and capacity projects, frequently alluded to as coordinated energy frameworks, improve the collaboration between various advancements.

Coordinating sun based or wind with energy capacity advances considers the persistent conveyance of power, beating the fluctuation innate in individual sustainable sources. These coordinated frameworks add to network soundness and backing the progress to a more decentralized and feasible energy foundation.

Chapter 5

Environmental Impacts and Remediation

Ecological effects and remediation are basic parts of our cutting edge world, where human exercises apply significant impacts on the common habitat. As our populace develops and businesses grow, the repercussions of these exercises on environments become progressively evident. This exposition investigates the diverse natural effects of human activities and looks at different remediation procedures pointed toward alleviating these impacts.

One of the essential natural effects originates from modern exercises, which discharge contaminations out of sight, water, and soil. Air contamination, a result of burning cycles in businesses and transportation, presents a horde of unsafe substances into the environment. The arrival of carbon dioxide (CO_2), sulfur dioxide (SO_2), nitrogen oxides (NO_x), and particulate matter adds to environmental change, corrosive downpour, and respiratory ailments. The results of these outflows are broad, influencing both human wellbeing and the uprightness of environments.

Notwithstanding air contamination, water contamination is a squeezing concern. Modern releases, horticultural spillover, and inappropriate garbage removal defile water bodies with poisonous substances, weighty metals, and supplements. The implications of water contamination stretch out past amphibian biological systems, influencing human populaces that depend on polluted water hotspots for drinking and agribusiness. The gathering of poisons in amphibian frameworks upsets the equilibrium of environments, prompting decreases in fish populaces, loss of biodiversity, and the rise of no man's lands absent any trace of oxygen.

Besides, soil contamination represents a critical danger to the climate and human prosperity. Pesticides, composts, and modern synthetics invade the dirt, corrupting its quality and fruitfulness. This tainting influences the soundness of plants, creatures, and people. Diligent natural contaminations (POPs), like pesticides and herbicides, can collect in the established pecking order, presenting long haul dangers to human wellbeing. Soil disintegration, exacerbated by deforestation and inappropriate land use, further escalates the corruption of soil quality and adds to the deficiency of arable land.

Deforestation, driven by the development of farming and logging, is a significant supporter of living space misfortune and biodiversity decline. Woods biological systems, which assume a significant part in carbon sequestration and keeping up with natural equilibrium, face extreme dangers from human exercises. The annihilation of woodlands reduces biodiversity as well as deliveries put away carbon into the environment, compounding environmental change. Preservation endeavors and maintainable land the board rehearses are fundamental for alleviating the effects of deforestation and safeguarding the natural elements of timberlands.

Environmental change, a worldwide peculiarity coming about because of the collection of ozone harming substances in the climate, is quite possibly of the most unavoidable natural test. Human exercises, especially the consuming of non-renewable energy sources, deforestation, and modern cycles, fundamentally add to the ascent in ozone depleting substance focuses. The results of environmental change incorporate climbing temperatures, modified precipitation designs, ocean level ascent, and more successive and serious outrageous climate occasions. These progressions present dangers to biological systems, farming, water assets, and human settlements, requiring earnest and deliberate endeavors to relieve and adjust to the effects of environmental change.

In light of these ecological difficulties, different remediation systems have been created to limit the adverse consequences of human exercises on the climate. Contamination control measures, for example, the establishment of emanation control gadgets in enterprises and the implementation of rigid natural guidelines, expect to diminish the arrival of poisons out of sight and water. The turn of events and reception of cleaner advancements, sustainable power sources, and energy-effective practices assume a critical part in relieving air contamination and tending to environmental change.

Wastewater treatment plants and the execution of best administration rehearses in horticulture are fundamental parts of water contamination remediation. By treating modern and metropolitan wastewater before release, impurities can be eliminated, guaranteeing that water bodies stay liberated from unsafe substances. Moreover, the advancement of economical agribusiness rehearses, for example, accuracy cultivating and natural cultivating, limits the utilization of substance inputs, decrease overflow, and safeguard water quality.

Soil remediation includes the expulsion or balance of impurities from the dirt to reestablish its wellbeing and ripeness. Methods, for example, phytoremediation, which utilizations plants to remove, amass, or corrupt contaminations, and bioremediation, which depends on microorganisms to separate pollutants, are acquiring unmistakable quality. These harmless to the ecosystem approaches offer feasible answers for soil contamination, keeping away from the requirement for broad removal and removal of debased soil.

Reforestation and afforestation programs add to the reclamation of corrupted scenes and the protection of biodiversity. Establishing local trees and carrying out supportable ranger service rehearses assist with modifying woods biological systems, upgrade carbon sequestration, and safeguard watersheds. Preservation drives, for example, the foundation of safeguarded regions and the authorization of against poaching measures, expect to save biodiversity and keep up with the environmental equilibrium of normal territories.

Relieving environmental change requires a complete methodology that incorporates decreasing ozone depleting substance outflows, improving carbon sequestration, and adjusting to the evolving environment. Progressing to environmentally friendly power sources, further developing energy effectiveness, and advancing supportable land use rehearses are fundamental parts of environmental change moderation. Afforestation and reforestation endeavors add to carbon sequestration, while feasible metropolitan preparation and versatile foundation assist networks with adjusting to the effects of an evolving environment.

Instructive projects and public mindfulness crusades assume an imperative part in encouraging natural stewardship and advancing reasonable practices. By bringing issues to light about the results of human exercises on the climate, people and networks can pursue informed decisions that add to ecological preservation. Natural training in schools, local area commitment drives, and the scattering of data through different media channels engage individuals to become advocates for a supportable and ecologically cognizant way of life.

Worldwide participation is urgent in tending to worldwide natural difficulties. Transboundary contamination, environmental change, and the deficiency of biodiversity require cooperative endeavors among countries to create and execute powerful arrangements.

Peaceful accords and deals, for example, the Paris Settlement on environmental change and the Show on Natural Variety, give structures to participation and coordination in resolving squeezing ecological issues. Conciliatory endeavors and shared liability are fundamental in accomplishing significant advancement toward a feasible and versatile future.

Notwithstanding the headway put forth in ecological remediation and protection attempts, various difficulties persevere. Proceeded with populace development, urbanization, and industrialization put extra tensions on the climate. Arising contaminations, like drugs and microplastics, present new difficulties that require imaginative arrangements. The interconnected idea of ecological issues requires all encompassing methodologies that think about the mind boggling cooperations between biological systems, human exercises, and the worldwide environment.

All in all, ecological effects and remediation are essential parts of our aggregate liability to shield the planet for people in the future. The outcomes of human exercises on air, water, soil, and biological systems feature the earnestness of taking on supportable practices and carrying out compelling remediation procedures. By embracing cleaner innovations, advancing protection endeavors, and encouraging global cooperation, we can pursue an agreeable conjunction with the regular world. As stewards of the Earth, it is occupant upon us to take a stab at a harmony between human turn of events and natural safeguarding, guaranteeing an economical and strong future for all.

5.1 Discuss the environmental consequences of coal mining and usage.

Coal mining and use have significant and sweeping ecological results that stretch out from the extraction of coal to its burning for energy creation. The ecological effects are multi-layered, influencing air, water, soil, and environments. This article dives into the unpredictable snare of ecological outcomes related with coal mining and use, investigating the difficulties and expected cures.

The extraction of coal through mining tasks represents a huge danger to the climate. Surface mining, including peak expulsion and strip mining, includes the evacuation of a lot of overlying stone and soil to get to coal creases. This cycle changes the geography of the land as well as prompts the annihilation of biological systems and living spaces. The dislodging of soil and rock can bring about avalanches, influencing downstream water quality and sea-going biological systems. Also, the evacuation of vegetation during surface mining upsets neighborhood environments, adding to living space misfortune and the downfall of biodiversity.

Underground coal mining, while less noticeably problematic than surface mining, has its own arrangement of ecological difficulties. Subsidence, the sinking of the World's surface because of the breakdown of underground mines, can prompt the adjustment of scenes and the disturbance of surface water stream.

The formation of mine voids can likewise present dangers of land precariousness, affecting framework and presenting threats to human networks. Besides, the extraction cycle discharges methane, a powerful ozone depleting substance, into the air. Methane outflows from coal mineshafts add to environmental change, worsening the ecological effect of coal mining.

Water contamination is an inescapable result of coal mining, influencing both surface water and groundwater. Corrosive mine seepage (AMD), a typical issue in regions with coal mining exercises, happens when sulfide minerals in coal stores respond with air and water, delivering sulfuric corrosive. This acidic overflow can taint close by water bodies, prompting the fermentation of streams and streams. The arrival of weighty metals, like iron, aluminum, and manganese, further debases water quality and stances endangers to amphibian life. Corrosive mine seepage hurts the prompt environment as well as make downstream impacts, influencing bigger stream frameworks and compromising water assets.

Coal mining activities frequently require significant measures of water for different cycles, including coal washing and residue concealment. The withdrawal of water from neighborhood water sources can prompt decreased water accessibility for different purposes, like farming and civil stockpile. In locales where water shortage is now a worry, the extra interest from coal digging fuels the opposition for restricted water assets. Moreover, the release of water from mining exercises, while possibly not appropriately made due, can present residue, contaminations, and raised temperatures into water bodies, further compromising sea-going biological systems.

The burning of coal for energy creation is a significant supporter of air contamination. Coal contains different pollutants, including sulfur, nitrogen, and minor components, which are delivered into the environment during burning. The ignition cycle produces sulfur dioxide (SO_2), nitrogen oxides (NO_x), particulate matter, and mercury emanations. These contaminations have huge ramifications for human wellbeing and the climate. Sulfur dioxide can prompt the arrangement of corrosive downpour, affecting soil and water quality and hurting oceanic life. Nitrogen oxides add to the development of ground-level ozone and fine particulate matter, which antagonistically affect respiratory wellbeing and can prompt brown haze arrangement.

Particulate matter discharged from coal burning postures dangers to both human wellbeing and the climate. Fine particulate matter, known as PM2.5, can infiltrate profound into the respiratory framework, causing respiratory and cardiovascular illnesses. Notwithstanding the wellbeing chances, particulate matter can store onto soil and water surfaces, influencing environments and adding to the eutrophication of water bodies. Mercury discharges from coal burning, as methylmercury, can bioaccumulate in sea-going creatures and posture dangers to human wellbeing through the utilization of sullied fish.

The ignition of coal is a significant wellspring of ozone harming substance outflows, fundamentally carbon dioxide (CO_2), which adds to worldwide environmental change. The arrival of CO_2 from consuming petroleum derivatives upgrades the nursery impact, prompting an expansion in worldwide temperatures. Environmental change has far reaching ecological results, including adjusted precipitation designs, more continuous and serious heatwaves, rising ocean levels, and disturbances to biological systems. The effects of environmental change, exacerbated by coal-terminated power plants and different wellsprings of fossil fuel byproducts, present dangers to biodiversity, food security, and the general steadiness of the World's environment framework.

Coal debris, a side-effect of coal ignition, represents extra natural difficulties. Coal debris contains different poisonous substances, including weighty metals like arsenic, lead, and mercury. Ill-advised removal and the board of coal debris, frequently put away in enormous impoundments or landfills, can prompt the draining of impurities into groundwater and surface water. Spills and breaks of coal debris storerooms have brought about disastrous natural calamities, making broad harm sea-going biological systems and presenting dangers to human wellbeing.

To address the ecological outcomes of coal mining and utilization, different remediation and relief methodologies are being investigated. Manageable mining rehearses, for example, recovery and reclamation of mined regions, mean to limit the drawn out effects of coal extraction. Revegetation endeavors assume a critical part in reestablishing environments impacted by surface mining, assisting with balancing out soils, forestall disintegration, and advance biodiversity. Propels in mining advances, including the improvement of cleaner extraction strategies and the utilization of computerization, add to lessening the natural impression of coal mining tasks.

The execution of water the executives rehearses is fundamental to moderate the water-related effects of coal mining. Treatment of corrosive mine seepage, through procedures, for example, lime dosing or built wetlands, can assist with killing acidic spillover and lessen the arrival of weighty metals into water bodies. Feasible water use works on, including water reusing and the utilization of elective water sources, add to limiting the freshwater interest of coal mining tasks. Also, administrative structures that implement severe water quality guidelines and require legitimate wastewater treatment are significant for forestalling and relieving water contamination.

In the domain of air quality, progressions in contamination control advancements are imperative for decreasing emanations from coal-terminated power plants. The establishment of vent gas desulfurization frameworks can catch sulfur dioxide, while specific reactant decrease and different advances target nitrogen oxide emanations.

Baghouse channels and electrostatic precipitators are utilized to catch particulate matter, forestalling its delivery into the environment. Past contamination control innovations, progressing to cleaner energy sources, like sustainable power and petroleum gas, is a central methodology for lessening the natural effect of coal burning.

Endeavors to address environmental change include a progress away from coal and other petroleum derivatives toward sustainable power sources. The development of environmentally friendly power framework, including sunlight based, wind, and hydropower, adds to decarbonizing the energy area and lessening ozone harming substance emanations. Strategies and impetuses that advance the reception of clean energy advances and stage out coal-terminated power plants assume a significant part in accomplishing the essential changes in the energy scene. Moreover, carbon catch and capacity (CCS) innovations, which catch CO2 discharges from coal-terminated power plants and store them underground, offer a likely road for relieving the environment effect of coal use.

Legitimate administration and removal of coal debris are basic parts of relieving the natural dangers related with this result. The improvement of elective purposes for coal debris, like in development materials or as a part in concrete, can lessen the volume of debris requiring removal and limit ecological dangers. Severe guidelines and checking systems for coal debris storage spaces are basic to forestall spills and holes, protecting close by environments and networks. Investigation into creative advancements for treating and settling coal debris can additionally upgrade the ecological manageability of coal burning side-effects.

Public mindfulness and local area commitment are fundamental for cultivating capable practices and affecting arrangement choices connected with coal mining and use. Instructing people group about the natural effects of coal and the significance of progressing to cleaner energy sources engage people to advocate for maintainable other options. Informed public talk can drive strategy changes, support the improvement of sustainable power framework, and add to a more extensive cultural shift towards additional economical and harmless to the ecosystem rehearses.

5.2 Explore efforts to remediate and rehabilitate areas affected by coal mining.

Endeavors to remediate and restore regions impacted by coal mining are fundamental parts of ecological stewardship and reasonable land use rehearses. The effects of coal mining, going from natural surroundings annihilation to water contamination and scene change, require complete methodologies for reestablishing biological systems, advancing biodiversity, and alleviating long haul ecological harm. This article dives into the different methodologies and drives pointed toward remediating and restoring regions impacted by coal mining, featuring the provokes and accomplishments in reestablishing harmony to these environments.

One of the essential difficulties in remediating regions affected by coal mining is the actual adjustment of scenes. Surface mining rehearses, for example, mountain ridge expulsion and strip mining, leave immense fields of land absent any trace of vegetation, with changed geography and disturbed biological systems. To address this, recovery endeavors center around reestablishing the land to a condition that upholds supportable environments and human use. Methods, for example, regrading, soil recreation, and revegetation assume vital parts in returning mined regions to a similarity to their normal state.

Regrading includes reshaping the land to its inexact unique form, relieving the extreme geological changes brought about by mining exercises. This interaction lessens the gamble of soil disintegration, advances regular seepage examples, and improves the general soundness of the recovered region. Via cautiously chiseling the territory, regrading adds to the tasteful and utilitarian reclamation of the scene, considering the re-foundation of biological systems and working with land utilize viable with the general climate.

Soil reproduction is a urgent part of recovering mined regions, as mining tasks frequently strip away fruitful dirt. Reestablishing soil fruitfulness and construction includes consolidating natural matter, supplements, and different changes to make a substrate helpful for plant development. This interaction not just backings the re-foundation of local vegetation yet additionally advances the recuperation of soil microbial networks fundamental for biological system working. Soil recreation means to imitate the piece and qualities of normal soils, cultivating a helpful climate for plant colonization and development.

Revegetation endeavors are necessary to reestablishing biodiversity and biological system capability in mined regions. Local plant species are painstakingly chosen and planted to speed up the recuperation of vegetation cover. The decision of plant species considers the natural qualities of the locale, including soil type, environment, and hydrology. Local plants assume a pivotal part in settling soils, forestalling disintegration, and giving territory to untamed life. Effective revegetation endeavors add to the general rebuilding of environment construction and capability, supporting the recuperation of the normal equilibrium upset by coal mining exercises.

Now and again, inventive methodologies, for example, the utilization of soil alterations and the use of mycorrhizal parasites are utilized to upgrade the progress of revegetation endeavors. Soil alterations, including lime, gypsum, and natural matter, assist with further developing soil design and supplement content, making better circumstances for plant foundation. Mycorrhizal parasites structure harmonious associations with plant roots, working with supplement take-up and advancing plant development. The utilization of these organisms can upgrade the flexibility of restored environments, especially in regions with testing soil conditions.

Wetland reclamation is one more basic part of coal mining remediation, as mining exercises frequently influence normal water stream designs and debase wetland biological systems. Developed wetlands, intended to impersonate the elements of regular wetlands, are made to channel and treat water releases from mining tasks.

These designed wetlands assist with eliminating toxins, silt, and overabundance supplements, further developing water quality before it enters downstream water bodies. Wetland reclamation contributes not exclusively to water remediation yet in addition to the re-foundation of special plant and creature networks adjusted to wetland environments.

Past the actual parts of remediation, endeavors are made to resolve the issue of corrosive mine seepage (AMD), an unavoidable issue in regions with coal mining exercises. AMD happens when sulfide minerals in coal stores respond with air and water, delivering sulfuric corrosive. This acidic overflow can debase water bodies, prompting the fermentation of streams and waterways. To remediate AMD, different methodologies are utilized, including the use of antacid substances, the utilization of wetland frameworks, and the execution of inactive treatment innovations.

The utilization of antacid substances, for example, limestone or lime, kills acridity in AMD-impacted waters. By expanding the pH of the water, these substances hasten metals and decrease the hurtful impacts of fermentation. The expansion of soluble materials can be applied straightforwardly to streams or integrated into treatment frameworks, giving a viable method for alleviating the effects of corrosive mine waste on sea-going environments.

Developed wetlands intended for AMD treatment use regular cycles to kill acridity and eliminate impurities. These wetlands are designed to support the development of plants that can endure acidic circumstances and work with the precipitation of metals. As water goes through these wetlands, the vegetation and microbial networks assume a urgent part in changing and immobilizing contaminations, further developing water quality downstream.

Uninvolved treatment advances, like penetrable responsive boundaries, are intended to catch and treat tainted groundwater or surface water. These boundaries contain responsive materials that kill causticity and eliminate metals as water moves through them. Inactive treatment frameworks offer a practical and supportable way to deal with tending to corrosive mine seepage, requiring insignificant upkeep once carried out.

Notwithstanding water-related difficulties, subsidence, a typical issue in regions with underground coal mining, presents dangers to the soundness of scenes and framework. Subsidence happens when underground voids made by mining exercises breakdown, prompting the sinking of the World's surface. To address subsidence, inlaying procedures include making up for the mined shortfalls with idle materials to balance out the land and forestall further sinking. This approach mitigates the physical and dangers related with subsidence, saving the trustworthiness of scenes and shielding human networks.

While recovery and restoration endeavors contribute essentially to relieving the natural effects of coal mining, the progress of these drives relies upon cautious preparation, checking, and versatile administration.

Long haul observing projects survey the advancement of recovered regions, following changes in vegetation, soil quality, and water quality. These checking endeavors give important information to assessing the adequacy of remediation methodologies and illuminating versatile administration practices to address arising difficulties.

Local area commitment and cooperation with partners are fundamental parts of fruitful remediation projects. Including neighborhood networks in the dynamic cycle and consolidating conventional natural information add to more viable and socially delicate remediation endeavors. Connecting with networks likewise cultivates a feeling of pride and obligation regarding the restored regions, advancing maintainable land use rehearses and guaranteeing the drawn out progress of remediation drives.

Administrative structures assume a significant part in directing and upholding recovery and restoration endeavors. Severe guidelines that command appropriate mine conclusion plans, monetary confirmations for recovery, and consistence with ecological principles are fundamental for considering digging organizations responsible for the natural effects of their activities. Administrative oversight guarantees that mining exercises stick to best practices and that sufficient measures are set up to address ecological difficulties all through the mining lifecycle.

The outcome of remediation endeavors is many times dependent upon the cooperation between government organizations, non-administrative associations, scholarly foundations, and the confidential area. Organizations and joint efforts influence different aptitude, assets, and points of view to foster imaginative and successful answers for the perplexing difficulties presented by coal mining influences. Drives that unite different partners cultivate an all encompassing and incorporated way to deal with remediation, tending to natural, social, and monetary parts of ecological reclamation.

In spite of the advancement in remediating regions impacted by coal mining, various difficulties persevere. Heritage mines, deserted before the execution of current ecological guidelines, keep on presenting dangers to biological systems and networks. The absence of monetary assets for recovery and the shortfall of people in question to address these heritage mines add to progressing natural corruption. Methodologies for tending to heritage mines incorporate focusing on high-risk locales, getting subsidizing for remediation, and carrying out extensive conclusion plans.

Arising impurities, for example, coal crease gas extraction side-effects, present new difficulties for remediation endeavors. The arrival of methane and other unpredictable natural mixtures from coal crease gas tasks can influence air quality and add to environmental change. Tending to these difficulties requires the advancement of creative innovations, administrative systems, and observing projects to alleviate the natural and wellbeing gambles related with arising impurities.

In endeavors to remediate and restore regions impacted by coal mining address a urgent move toward moderating the natural effects of mining exercises. Recovery works on, including regrading, soil reproduction, and revegetation, mean to reestablish environments and advance manageable land use. Remediation systems for water-related difficulties, for example, corrosive mine waste, consolidate methods like the use of antacid substances, developed wetlands, and detached treatment innovations. Addressing subsidence chances includes refilling mined voids to balance out scenes and safeguard framework.

Long haul observing, local area commitment, and administrative structures are basic parts of effective remediation projects, guaranteeing responsibility, versatile administration, and the fuse of nearby information. Cooperation among partners, including government organizations, NGOs, the scholarly world, and the confidential area, upgrades the adequacy and maintainability of remediation endeavors. Regardless of the headway made, difficulties, for example, heritage mines and arising pollutants highlight the requirement for progressing exploration, development, and a purposeful worldwide work to address the complex ecological results of coal mining. Through these aggregate undertakings, we can endeavor towards a more manageable and tough conjunction with the scenes impacted by coal mining, cultivating a harmony between human turn of events and ecological protection.

5.3 Successful environmental restoration projects.

Fruitful natural reclamation projects embody mankind's ability to recuperate and rejuvenate biological systems that have been debased by different anthropogenic exercises. These ventures length assorted scenes and biological systems, exhibiting the flexibility of rebuilding ways to deal with various settings. This paper investigates striking instances of effective natural rebuilding projects, featuring the methodologies utilized, the positive results accomplished, and the illustrations learned for future undertakings.

1. **The Loess Level Watershed Restoration Undertaking, China:**

   One of the most broad and aggressive ecological rebuilding projects all around the world is the Loess Level Watershed Restoration Undertaking in China. Started in the last part of the 1990s, this venture expected to address serious soil disintegration and land corruption on the Loess Level, a region that had encountered hundreds of years of impractical farming practices. The task utilized an all encompassing methodology, incorporating soil and water protection measures, afforestation, and reasonable land the board rehearses.

   Key systems incorporated the development of really look at dams and porches to control water overflow and diminish soil disintegration. Huge scope afforestation endeavors included establishing local vegetation to settle the dirt and forestall further debasement. Also, reasonable rural practices, for example, shape furrowing and agroforestry, were acquainted with advance soil preservation while supporting neighborhood vocations.

   The results of the Loess Level undertaking have been surprising. Soil disintegration has been essentially diminished, and vegetation cover has expanded, prompting further developed water quality and improved biodiversity. The venture has reestablished the environmental equilibrium as well as had positive financial effects, with neighborhood networks profiting from expanded rural efficiency and worked on day to day environments.

2. **The Everglades Rebuilding Undertaking, US:**

   The Everglades, a remarkable and biologically significant wetland environment in Florida, confronted critical debasement because of many years of water the executives practices, urbanization, and farming turn of events. The Thorough Everglades Reclamation Plan (CERP), started in 2000, addresses a complete and cooperative work to reestablish the normal hydrological examples of the Everglades and address biological system decline.

   CERP incorporates a large number of reclamation techniques, for example, the change of water framework to imitate normal water stream, the evacuation of obtrusive species, and the making of water stockpiling and treatment regions. By once again introducing normal water stream designs, the venture plans to reestablish territories, further develop water quality, and restore the noteworthy harmony among freshwater and saltwater in the locale.

While the Everglades Reclamation Venture is as yet continuous, there have been eminent victories. Rebuilding endeavors in specific regions have brought about superior water quality, expanded populaces of local species, and the arrival of some normal hydrological processes. The task underscores the significance of versatile administration, long haul responsibility, and interdisciplinary joint effort in tending to complex environmental difficulties.

3. **The Aral Ocean Reclamation Venture, Focal Asia:**
The Aral Ocean, when the fourth-biggest lake on the planet, confronted disastrous downfall because of unreasonable water use for water system in the encompassing district. The redirection of waterways that took care of into the Aral Ocean prompted its shrinkage and the rise of natural and financial emergencies in the encompassing networks. In light of this environmental catastrophe, the World Bank, as a team with Focal Asian nations, started the Aral Ocean Bowl Program in the mid 2000s.

The undertaking zeroed in on further developing water the board works on, advancing supportable farming, and reestablishing the biological equilibrium of the Aral Ocean. Key mediations incorporated the development of water-saving designs, the presentation of water-productive farming practices, and the restoration of a North Aral Ocean through the development of a dam. The fruitful endeavors in the North Aral Ocean have prompted expanded water levels and worked on neighborhood biological systems.

While challenges endure, and the full rebuilding of the whole Aral Ocean stays a great undertaking, the advancement in the North Aral Ocean exhibits the potential for cooperative, transboundary endeavors to address enormous scope ecological corruption.

4. **The Chesapeake Cove Rebuilding, US:**
The Chesapeake Sound, the biggest estuary in the US, confronted extreme water quality issues because of supplement contamination, sedimentation, and the effect of urbanization and agribusiness in the watershed. The Chesapeake Narrows Program, started in 1983, addresses an extensive, cooperative exertion including various states, government organizations, and nearby partners to reestablish the strength of the cove.

The reclamation plan centers around lessening supplement spillover, working on agrarian practices, and executing stormwater the executives measures. Endeavors incorporate the foundation of riparian cushions, cover crops, and the redesign of wastewater treatment plants. The execution of best administration rehearses and administrative measures has intended to decrease supplement and residue inputs into the sound.

Throughout the long term, the Chesapeake Sound rebuilding project has seen positive results. Water quality has improved, and the overflow of submerged grasses, a critical mark of biological system wellbeing, has expanded. The task highlights the significance of composed, science-based approaches, and the commitment of different partners in tending to complex natural difficulties.

5. **The Rewilding of Oostvaardersplassen, Netherlands:**
The Oostvaardersplassen in the Netherlands addresses a novel examination in rewilding — a type of ecological rebuilding that permits regular cycles to shape environments. Initially made as a polder for modern purposes, the region was subsequently assigned as a nature save during the 1960s. The rewilding approach in Oostvaardersplassen includes negligible human mediation, permitting the scene to advance through normal natural cycles.

Enormous herbivores, including Konik ponies and red deer, were acquainted with mirror the job of wiped out wild herbivores. The idea is to make a self-supporting environment where regular cycles like predation, herbivory, and vegetation elements shape the scene. The task has stood out for its creative way to deal with reestablishing environments without broad human administration.

While questionable and dependent upon progressing banter, the rewilding of Oostvaardersplassen has shown the potential for nature to bounce back whenever offered the chance. It brings up issues about the harmony among mediation and non-intercession in reclamation projects and the job of wild in contemporary preservation endeavors.

6. **The Yaeda Valley People group Drove Preservation, Tanzania:**
In the Yaeda Valley of northern Tanzania, the Hadza public, one of the final agrarian networks, have embraced local area drove preservation endeavors to safeguard their conventional terrains and save biodiversity. Confronting dangers from land infringement, logging, and agrarian extension, the Hadza public, with help from NGOs like Carbon Tanzania, have carried out local area based preservation drives.

The Hadza public have laid out local area drove preservation regions where customary land use rehearses, like controlled consumes and supportable hunting, are utilized to keep up with biological system wellbeing. Also, carbon offset projects have been created, connecting the protection of woodlands with worldwide endeavors to alleviate environmental change. The Hadza's endeavors embody the significance of nearby information, local area strengthening, and the reconciliation of customary practices into present day protection methodologies.

7. **The Kihansi Chasm Hydroelectric Dam Expulsion, Tanzania:**
In the Kihansi Chasm of Tanzania, the development of a hydroelectric dam essentially affected the neighborhood environment, especially the Kihansi shower frog, an animal categories endemic to the locale. The dam modified the normal water stream and made conditions troublesome for the frog's living space, prompting an extraordinary decrease in its populace.

Accordingly, an exceptional and cooperative exertion was started to eliminate the dam and reestablish the normal progression of the Kihansi Stream. The evacuation of the dam permitted the rebuilding of the amphibian's natural surroundings, and resulting renewed introduction endeavors have prompted the recuperation of the Kihansi shower frog populace. The undertaking grandstands the potential for designated mediations to turn around the effects of explicit natural aggravations and the significance of versatile administration in protection.

8. **The Bonneville Natural Establishment's Water Reclamation Authentications, US:**

The Bonneville Natural Establishment (BEF) in the US spearheaded the idea of Water Reclamation Endorsements (WRCs) to address the environmental effects of water utilization. WRCs address a market-based way to deal with water rebuilding, permitting people and organizations to buy testaments that asset projects pointed toward reestablishing water to exhausted biological systems.

BEF puts the returns from WRC deals in tasks, for example, streamflow reclamation, living space upgrade, and water quality improvement. These ventures, executed as a team with neighborhood accomplices, add to the biological soundness of watersheds and give a system to people and organizations to balance their water impression. The WRC model delineates how creative funding instruments can uphold enormous scope natural rebuilding drives.

9. **The Mount St. Helens Environment Recuperation, US:**

The ejection of Mount St. Helens in 1980 caused broad obliteration, clearing out immense areas of woods and changing the scene emphatically. Soon after the emission, researchers and land supervisors noticed the regular course of environment recuperation, exhibiting the versatility of nature.

The shortfall of human mediation considered the unconstrained colonization of plant and creature species in the crushed regions. Analysts archived the continuous return of vegetation, the foundation of new soil, and the recolonization of untamed life. The Mount St. Helens recuperation fills in as a characteristic examination in biological progression, underlining the significance of permitting environments to recover through regular cycles.

10. **The Elwha Stream Reclamation, US:**

The Elwha Stream Reclamation Task in Washington state addresses one of the main dam expulsion endeavors ever. Two huge hydroelectric dams, the Elwha and Glines Gully Dams, were built on the Elwha Stream in the mid twentieth hundred years, hindering salmon relocation, adjusting silt transport, and affecting generally speaking waterway wellbeing.

In 2011 and 2014, the dams were eliminated in a fantastic work to reestablish the waterway environment. Dam evacuation permitted the arrival of salmon to their notable generating grounds, restored dregs transport, and rejuvenated riparian territories.

The Elwha Waterway Rebuilding is a demonstration of the expected advantages of dam expulsion in reestablishing stream environments and advancing the recuperation of cornerstone species.

Ecological rebuilding projects assume a vital part in relieving the effects of human exercises on environments, cultivating biodiversity, and advancing reasonable collaborations among nature and society. These ventures incorporate a large number of drives pointed toward restoring corrupted scenes, reestablishing biological system works, and upgrading the versatility of regular natural surroundings. This article investigates the assorted methodologies and eminent instances of natural rebuilding projects, stressing their significance in addressing ecological difficulties and adding to the more extensive objectives of preservation and manageability.

1. **The Incomparable Green Wall Drive, Africa:**
   The Incomparable Green Wall is a visionary undertaking spreading over different African nations, pointed toward battling desertification and land debasement in the Sahel district. Sent off in 2007, the drive looks to make a mosaic of green and useful scenes across the mainland, extending from Senegal in the west to Djibouti in the east. The task includes establishing a belt of dry spell safe trees, bushes, and grasses to forestall soil disintegration, further develop water maintenance, and backing nearby networks.

   This cross-country exertion resolves ecological issues as well as has financial ramifications. By giving elective vocations, for example, agroforestry and economical land the board rehearses, the Incomparable Green Wall engages neighborhood networks, improves food security, and adds to environmental change variation. While the undertaking is continuous, it represents the cooperative and long haul nature of enormous scope reclamation drives.

2. **The Serengeti Rebuilding Undertaking, Tanzania:**
   The Serengeti environment, eminent for its famous natural life and movement designs, confronted dangers from intrusive plant species and territory debasement. The Serengeti Reclamation Venture, started by the Frankfurt Zoological Society, centers around eliminating intrusive species, reestablishing prairies, and restoring regular fire systems. By once again introducing controlled consumes, the undertaking plans to revive the prairies, giving a favorable climate to local verdure.

   The rebuilding endeavors in the Serengeti support biodiversity as well as add to the safeguarding of basic territories for transient species. The undertaking features the significance of versatile administration, logical examination, and joint effort between preservation associations and nearby networks in accomplishing fruitful environment reclamation.

3. **The Eden Task, Joined Realm:**
   The Eden Venture in Cornwall, UK, remains as an imaginative illustration of

biological system rebuilding inside a metropolitan setting. At first a neglected china earth quarry, the site was changed into a progression of biomes, each recreating various environments and biological systems. The task means to exhibit the variety of vegetation, bring issues to light about ecological issues, and rouse reasonable living practices.

The Eden Venture shows the way that innovative plan and public commitment can add to both biological reclamation and ecological instruction. It fills in as a model for reusing corrupted scenes for sporting, instructive, and protection purposes, stressing the potential for metropolitan regions to become impetuses for positive ecological change.

4. **The Mount Grandiose Reaches Reclamation Program, Australia:**
Australia's Mount Grand Reaches confronted natural corruption because of land clearing, obtrusive species, and modified fire systems. The Mount Grand Reaches Reclamation Program, sent off by the Australian government, centers around reestablishing local vegetation, overseeing intrusive species, and executing fire the board systems. The task includes cooperation with nearby networks, landowners, and ecological associations to accomplish scene scale rebuilding.

Through essential preparation and local area inclusion, the reclamation program means to upgrade biodiversity, further develop water quality, and reestablish biological system strength. The task's accentuation on versatile administration and partner commitment gives significant experiences into tending to ecological difficulties in unique and different scenes.

5. **The Seas Without Boundaries Drive, Worldwide:**
Marine environments face various dangers, including overfishing, natural surroundings debasement, and environmental change. The Seas Without Lines drive, drove by the Untamed life Protection Society, centers around huge scope marine preservation and rebuilding endeavors across numerous nations. By laying out marine safeguarded regions, advancing supportable fishing practices, and directing examination on marine biodiversity, the drive looks to upgrade the wellbeing and versatility of marine environments.

The worldwide extent of Seas Without Lines perceives the interconnectedness of sea environments and the requirement for cooperative, cross-line protection techniques. It features the significance of worldwide participation in tending to the difficulties looked by marine conditions and stresses the job of science-based approaches in marine rebuilding.

6. **The Aichi Targets, Worldwide:**
The Aichi Targets, part of the Brilliant course of action for Biodiversity 2011-2020 under the Show on Organic Variety, put forth aggressive objectives for worldwide biodiversity preservation and rebuilding. These objectives incorporate tending to environment misfortune, forestalling the eradication of species, and reestablishing biological systems that have been debased. While

progress towards the Aichi Targets has been shifted, the system fills in as a complete aide for nations to create and carry out rebuilding methodologies custom-made to their one of a kind biological settings.

The Aichi Targets feature the significance of incorporating biodiversity preservation and rebuilding into public and global arrangement plans. They underline the job of safeguarded regions, supportable land the board, and the commitment of nearby networks in accomplishing worldwide protection objectives.

7. **The Rewilding of Yellowstone Public Park, US:**

The renewed introduction of dim wolves to Yellowstone Public Park during the 1990s is a milestone instance of rewilding — a methodology that expects to reestablish normal biological cycles by once again introducing cornerstone species. The shortfall of wolves had prompted overpopulation of herbivores, like elk, bringing about overgrazing and environment debasement. The renewed introduction of wolves had flowing consequences for the biological system, impacting elk conduct, diminishing overgrazing, and advancing the recuperation of plant networks.

The Yellowstone rewilding project highlights the significance of cornerstone species in keeping up with environment equilibrium and features the potential for trophic fountains to drive biological rebuilding. It has turned into a paradigmatic illustration of how the rebuilding of a solitary animal varieties can significantly affect the whole biological system.

8. **The Danube Waterway Reclamation Program, Europe:**

The Danube Stream, one of Europe's significant streams, confronted contamination, natural surroundings misfortune, and adjusted stream systems because of industrialization and urbanization. The Danube Stream Reclamation Program, drove by the Global Commission for the Insurance of the Danube Stream, centers around further developing water quality, reestablishing floodplains, and upgrading the environmental network of the waterway.

Through measures like the expulsion of outdated dams, the reclamation of wetlands, and the execution of water quality administration designs, the program expects to revive the Danube biological system. The cooperative endeavors of different nations along the waterway exhibit the significance of transboundary collaboration in accomplishing fruitful stream rebuilding.

9. **The Rainforest Establishment, Worldwide:**

The Rainforest Establishment, working universally with an emphasis on tropical rainforests, takes part in projects that expect to forestall deforestation, reestablish corrupted regions, and backing native networks. By joining preservation endeavors with reasonable improvement drives, the establishment tends to the mind boggling exchange between human livelihoods and biological system wellbeing.

Projects embraced by the Rainforest Establishment incorporate reforestation, agroforestry programs, and the foundation of local area oversaw protection regions. The establishment's comprehensive methodology perceives the vital job of neighborhood networks in preservation and underscores the requirement for enabling native people groups in the rebuilding and security of rainforests.

10. **The Thousand years Seed Bank Organization, Worldwide:**

The Thousand years Seed Bank Organization, drove by the Regal Botanic Nurseries, Kew, is a worldwide drive zeroed in on rationing and reestablishing plant variety through seed banking. The undertaking includes gathering, putting away, and exploring seeds from around the world, determined to defend plant species and supporting territory reclamation endeavors.

The seed bank's endeavors add to ex situ preservation, giving an asset to reestablishing environments, particularly in locales confronting territory misfortune and environmental change. By saving the hereditary variety of plant species, the Thousand years Seed Bank Organization assumes a basic part in improving the strength of biological systems to ecological stressors.

Chapter 6

Policies and Global Initiatives

In the quickly developing scene of worldwide relations, strategies and worldwide drives assume a urgent part in forming the course of countries and impacting the direction of worldwide turn of events. These strategies, created by legislatures, worldwide associations, and cooperative endeavors, act as the structure for tending to a heap of difficulties and open doors that rise above public boundaries. From financial collaboration to natural manageability, from civil rights to innovative headways, the approaches and drives concocted at the worldwide level convey significant ramifications for the interconnected universe of the 21st hundred years.

One of the general topics in contemporary worldwide arrangements spins around practical turn of events. The global local area has progressively perceived the basic of offsetting monetary development with natural stewardship and social value. The Unified Countries Supportable Improvement Objectives (SDGs) stand as a demonstration of this responsibility, giving an extensive diagram to resolving issues like destitution, hunger, wellbeing, instruction, orientation fairness, clean water, and environment activity. Legislatures and associations overall have adjusted their strategies to the SDGs, encouraging an aggregate work to construct an additional practical and comprehensive future.

Financial strategies, as well, assume a significant part in forming the worldwide scene. The elements of worldwide exchange, venture, and money are represented by a perplexing trap of strategies that look to adjust contending interests and guarantee fair cooperation in the worldwide economy. Economic alliance, like the World Exchange Association (WTO) arrangements or territorial exchange settlements like the European Association, shape the guidelines and guidelines that oversee the development of labor and products across borders. These arrangements expect to advance fair contest, forestall protectionism, and encourage financial development on a worldwide scale.

Lately, the ascent of computerized innovation has achieved another aspect to worldwide strategies. The quick speed of innovative progression has outperformed the capacity of customary administrative systems to keep up. Issues like information protection, network safety, and the moral ramifications of man-made reasoning have become focal worries for policymakers around the world. Drives like the Overall Information Security Guideline (GDPR) in the European Association and conversations inside worldwide discussions highlight the requirement for a blended way to deal with overseeing the computerized domain.

Worldwide wellbeing strategies have acquired uplifted significance, especially right after the Coronavirus pandemic. The interconnectedness of the world has made it basic for countries to team up on wellbeing drives, data sharing, and antibody circulation. Associations like the World Wellbeing Association (WHO) assume a focal part in planning worldwide reactions to wellbeing emergencies and forming rules to guarantee the prosperity of populaces across the globe. The pandemic has highlighted the requirement for hearty worldwide wellbeing foundation and participation to address current difficulties as well as to get ready for future wellbeing emergencies.

Natural strategies structure a basic part of the worldwide plan, given the rising dangers presented by environmental change and ecological debasement. Peaceful accords, for example, the Paris Understanding expect to unite countries in a deliberate work to relieve the effects of environmental change and progress towards a manageable, low-carbon future. The direness of addressing natural difficulties has prompted a developing accentuation on green strategies, sustainable power drives, and protection endeavors at the worldwide level.

In the domain of basic liberties and civil rights, worldwide drives try to make a reality where central freedoms are secured, and inclusivity is supported. Associations like Reprieve Worldwide and Basic freedoms Watch work to focus on denials of basic freedoms, while the Unified Countries Common liberties Board fills in as a discussion for countries to address and review infringement. Orientation balance has likewise turned into a focal concentration, with drives, for example, HeForShe planning to draw in men in the battle for ladies' freedoms and correspondence.

With regards to harmony and security, global approaches are intended to forestall clashes, resolve questions, and advance dependability. The Unified Countries Security Chamber assumes a focal part in tending to worldwide security issues, approving peacekeeping missions, and forcing sanctions when essential. Peace treaties, like the Settlement on the Limitation of Atomic Weapons (NPT), intend to forestall the spread of atomic weapons and advance demobilization. Counterterrorism endeavors, transnational wrongdoing collaboration, and peacebuilding drives all add to the perplexing embroidered artwork of worldwide security arrangements.

The monetary variations among created and emerging countries have prodded drives pointed toward advancing comprehensive development and decreasing neediness. Unfamiliar guide, advancement help, and obligation alleviation programs are fundamental parts of worldwide endeavors to address disparity and inspire minimized networks. The Assembled Countries Improvement Program (UNDP) and the World Bank are central participants in organizing global endeavors to advance practical turn of events and destroy neediness.

Relocation strategies have additionally acquired unmistakable quality as the development of individuals across borders keeps on molding socioeconomics and social scenes. The difficulties presented by constrained removal, evacuees, and refuge searchers have provoked the advancement of worldwide structures, like the Worldwide Conservative for Protected, Deliberate, and Standard Movement. Adjusting the necessities of transients with the worries of host countries stays a complicated strategy challenge, requiring a sensitive exchange of compassionate contemplations and public interests.

Notwithstanding arising worldwide difficulties, cooperative drives are progressively turning into the standard. Multilateralism, as exemplified by associations like the Unified Countries, looks to resolve gives that rise above public limits through aggregate activity. In any case, the adequacy of multilateral methodologies is dependent upon the eagerness of countries to participate and maintain the standards of shared liability.

The job of delicate power in molding worldwide discernments and impacting conduct has likewise turned into a vital thought in global strategies. Social tact, public discretion, and worldwide telecom are instruments utilized by countries to project their qualities, thoughts, and effect on the worldwide stage. Delicate power, as instituted by political specialist Joseph Nye, features the capacity of countries to accomplish their targets through fascination and influence instead of pressure.

The ascent of populism and patriotism in different regions of the planet has brought another dynamic into worldwide governmental issues. Countries wrestling with inner difficulties frequently turn internal, rethinking their responsibilities to peaceful accords and coalitions. The pressures between public power and worldwide participation represent a huge test to the viability of worldwide strategies, requiring cautious route and political artfulness.

In the domain of innovation, the administration of arising advances like man-made brainpower, biotechnology, and quantum figuring presents an impressive strategy challenge. As these advancements reshape enterprises and social orders, the requirement for moral rules, administrative systems, and global participation turns out to be progressively obvious. The mindful turn of events and organization of these innovations are urgent to guaranteeing that they benefit mankind in general and don't worsen existing imbalances.

The Unified Countries, as the transcendent worldwide association, fills in as a gathering for political discourse, compromise, and the detailing of worldwide strategies. Containing particular organizations, projects, and bodies, the UN tends to a wide exhibit of issues, from peacekeeping and improvement to wellbeing and common freedoms. The Overall Get together, the Security Committee, and the Global Courtroom are key parts of the UN's design, each assuming a particular part in keeping up with worldwide harmony and security.

Worldwide administration structures stretch out past the UN, including territorial associations and partnerships that address explicit difficulties inside their ranges of prominence. The European Association, the African Association, the Relationship of Southeast Asian Countries (ASEAN), and the Association of American States (OAS) are instances of local bodies that add to the plan and execution of strategies customized to the necessities of their separate districts.

In the financial domain, foundations like the Global Money related Asset (IMF) and the World Bank assume pivotal parts in giving monetary help, advancing financial dependability, and supporting advancement projects in part nations. These foundations additionally act as gatherings for worldwide monetary collaboration, where approaches are talked about, and rules are laid out to explore the intricacies of the worldwide economy.

Worldwide approaches and drives are not static; they should adjust to advancing difficulties and open doors. The course of strategy definition is intrinsically unique, including the persistent appraisal of international turns of events, financial patterns, innovative headways, and cultural changes. Adaptability and responsiveness are key ascribes of viable worldwide approaches, permitting them to resolve arising issues and profit by additional opportunities.

The viability of worldwide strategies relies upon the level of responsibility and consistence displayed by countries. While peaceful accords and settlements give a structure to participation, their prosperity depends on the readiness of part states to execute and stick to the settled upon measures. Requirement components, discretionary tension, and general assessment all assume parts in affecting the way of behaving of countries in the worldwide field.

The significance of public commitment and common society support in the detailing and execution of worldwide approaches couldn't possibly be more significant. In a time of expanded network and data dispersal, the voices of residents, non-administrative associations (NGOs), and grassroots developments have the ability to shape the talk around worldwide issues.

Online entertainment stages, specifically, have become integral assets for preparing general assessment and considering policymakers responsible.

As the world explores the intricacies of the 21st hundred years, the requirement for imaginative and forward-looking worldwide approaches turns out to be always evident. The interconnected idea of worldwide difficulties — from pandemics and environmental change to monetary disparity and innovative interruptions — requires a comprehensive and cooperative methodology. Policymakers should be skilled at expecting future patterns, utilizing innovative headways, and encouraging global collaboration to address the multi-layered difficulties within recent memory.

6.1 Analyze government policies promoting the transition from coal to cleaner energy.

The progress from coal to cleaner energy sources addresses a basic part of worldwide endeavors to address environmental change, decrease ecological effect, and cultivate feasible turn of events. States overall have perceived the direness of moving away from non-renewable energy sources, especially coal, because of its critical commitments to ozone harming substance discharges and air contamination. Dissecting the public authority strategies advancing this progress divulges an intricate interaction of financial, ecological, and social contemplations that shape the direction of energy changes.

At the core of government approaches advancing the shift from coal to cleaner energy is the basic to moderate environmental change. Coal burning is a significant wellspring of carbon dioxide (CO_2) emanations, contributing essentially to the ozone harming substance focuses in the air. Perceiving the job of these emanations in an unnatural weather change and its related effects, state run administrations have focused on peaceful accords, for example, the Paris Arrangement, which sets focuses for lessening discharges and restricting worldwide temperature increments. In arrangement with these responsibilities, numerous nations have planned approaches pointed toward progressively eliminating coal-terminated power plants and empowering the reception of cleaner options.

Financial contemplations likewise assume a significant part in molding government strategies connected with the progress from coal. The sustainable power area, including sun oriented, wind, and hydropower, has seen momentous mechanical headways and cost decreases throughout the long term. Legislatures perceive the financial open doors related with putting resources into clean energy advances, like work creation, development, and upgraded energy security. Policymakers frequently plan motivations, endowments, and administrative structures to advance the development of the environmentally friendly power industry, making it more serious and alluring to financial backers.

One normal strategy instrument for advancing the progress is the execution of environmentally friendly power targets. Legislatures put forth unambiguous objectives for the portion of environmentally friendly power in their general energy blend, flagging a promise to decreasing dependence on coal and other petroleum derivatives. These objectives act as a guide for the energy progress and give an unmistakable sign to financial backers, organizations, and general society in regards to the heading of the country's energy strategies. The European Association's Environmentally friendly power Mandate, for instance, lays out restricting focuses for the portion of environmentally friendly power in the all out energy utilization of part states.

Monetary impetuses and endowments are pivotal parts of government strategies advancing the progress from coal. Perceiving the higher introductory expenses related with environmentally friendly power projects, states frequently offer monetary help to make clean energy all the more monetarily feasible. This help can take different structures, including tax reductions, feed-in duties, awards, and low-interest advances. By decreasing the monetary hindrances to passage, these motivators empower interest in sustainable power foundation and add to the general seriousness of clean energy advances.

Administrative structures likewise assume a focal part in working with the change to cleaner energy. States carry out guidelines that set discharge norms, require the conclusion of maturing coal plants, and lay out a level battleground for clean energy innovations. Carbon valuing components, for example, carbon expenses or cap-and-exchange frameworks, give financial motivations to decreasing outflows by doling out an expense for carbon contamination. Very much planned guidelines establish a helpful climate for the development of sustainable power and boost ventures to embrace cleaner rehearses.

Government strategies advancing the change from coal to cleaner energy frequently include a staged methodology. Perceiving the current foundation and the monetary effect on networks subject to coal, policymakers foster systems to deal with the progress in a fair and impartial way. Simply Progress arrangements expect to help laborers and networks impacted by the decay of the coal business, giving retraining open doors, financial broadening, and social wellbeing nets. These strategies recognize the significance of guaranteeing that nobody is abandoned in the shift to a more feasible energy future.

Global participation and joint effort are indispensable parts of government arrangements advancing the progress from coal. Given the transboundary idea of ecological difficulties, countries frequently participate in cooperative endeavors to share best practices, advancements, and monetary assets. Drives like Mission Advancement, a worldwide organization of nations focused on multiplying their spotless energy innovative work ventures, embody the cooperative soul driving the change to cleaner energy on the global stage.

The advancement of development and innovative work (Research and development) is a vital technique in government strategies for progressing from coal. Putting resources into mechanical progressions is fundamental for making cleaner energy sources more proficient, reasonable, and adaptable. State run administrations frequently dispense assets for Research and development projects, lay out research habitats, and boost private-area interest in clean energy advancement. Leap forwards in energy capacity, framework reconciliation, and environmentally friendly power advances are basic for defeating the discontinuity and unwavering quality difficulties related with some cleaner energy sources.

Nearby and local states likewise assume an essential part in driving the progress from coal. As a rule, subnational elements have more straightforward cooperations with impacted networks and ventures, taking into consideration custom fitted arrangements that address neighborhood challenges. Neighborhood states might execute drafting guidelines, energy effectiveness projects, and local area sun powered drives to speed up the reception of cleaner energy at the grassroots level.

Social acknowledgment and public commitment are huge variables that impact the outcome of government arrangements advancing the progress from coal. State run administrations perceive the significance of encouraging public help for clean energy drives, as open discernment can shape the political will to carry out and support these arrangements. Correspondence techniques, public mindfulness missions, and local area counsels are utilized to instruct people in general about the advantages of changing to cleaner energy and to address concerns connected with work relocation and monetary effects.

Difficulties and intricacies have large amounts of the execution of government strategies advancing the progress from coal. The energy progress includes an essential rebuilding of energy frameworks, with suggestions for different partners, including energy makers, laborers, and networks. The transitioning away from of coal can prompt financial separation in locales vigorously subject to coal mining and power age. Policymakers should cautiously explore the social and financial components of the progress, adjusting the basic of ecological supportability with the requirement for a fair and comprehensive interaction.

One more test is the energy framework's dormancy, described by existing foundation, personal stakes, and laid out methods of energy creation. Defeating this idleness requires strategy mediations as well as the improvement of a versatile and adaptable energy framework fit for coordinating new innovations and adjusting to evolving conditions. State run administrations should address matrix joining difficulties, stockpiling advances, and the improvement of brilliant networks to guarantee the unwavering quality and soundness of the energy supply during the change.

The international affairs of energy progress likewise present difficulties, as countries look to tie down admittance to basic assets for clean energy advancements. The creation and supply chains of sustainable power parts, like intriguing earth minerals for batteries and photovoltaic cells, present new contemplations for worldwide collaboration and rivalry. States should explore these intricacies to guarantee a safe and economical stock of assets for the perfect energy progress.

All in all, administration strategies advancing the change from coal to cleaner energy mirror a complex methodology that considers monetary, ecological, and social elements. The basic to moderate environmental change, combined with the financial open doors related with clean energy advancements, drives the detailing of strategies that boost the reception of sustainable power sources. Sustainable power targets, monetary impetuses, administrative structures, and simply progress arrangements structure necessary parts of these methodologies.

The change from coal to cleaner energy isn't without difficulties, and policymakers should explore issues connected with monetary separation, energy framework dormancy, and international contemplations. Cooperation at the global, public, and neighborhood levels, combined with a promise to development and public commitment, is vital for conquering these difficulties and understanding a practical energy future. As states overall keep on wrestling with the intricacies of energy advances, the viability of their strategies will shape the direction of worldwide endeavors to address environmental change and fabricate a more practical and versatile energy biological system.

6.2 Explore international collaborations and agreements addressing coal-related issues.

Worldwide joint efforts and arrangements tending to coal-related issues are crucial parts of the worldwide reaction to ecological difficulties, environmental change, and maintainable turn of events. Coal, as a predominant wellspring of energy for quite a long time, has been related with different ecological and social issues, including air contamination, ozone harming substance emanations, and unfavorable wellbeing impacts. Perceiving the cross-line nature of these difficulties, countries have taken part in cooperative endeavors to alleviate the effects of coal and progress toward cleaner, more reasonable energy sources. Inspecting the scene of worldwide joint efforts and arrangements gives bits of knowledge into the complicated trap of discretionary, ecological, and financial contemplations that shape the worldwide reaction to coal-related issues.

One of the most striking peaceful accords tending to coal-related issues is the Paris Understanding. Taken on in 2015 under the Unified Countries System Show on Environmental Change (UNFCCC), the Paris Understanding addresses a milestone accord pointed toward restricting worldwide temperature increments and relieving the effects of environmental change.

While not expressly centered around coal, the arrangement recognizes the need to change away from petroleum derivatives, including coal, to accomplish the settled upon environment objectives.

Under the Paris Arrangement, nations submit broadly resolved commitments (NDCs), illustrating their singular focuses for lessening ozone depleting substance outflows. Numerous countries incorporate procedures for progressively transitioning away from coal and expanding the portion of environmentally friendly power in their NDCs. The cooperative idea of the arrangement urges nations to share best practices, advances, and monetary assets to help the progress to cleaner energy sources.

Inside the system of the Paris Understanding, the Controlling Past Coal Partnership arose as a cooperative drive. Sent off in 2017 by Canada and the Assembled Realm, the collusion unites state run administrations, organizations, and associations focused on transitioning away from unabated coal power and supporting impacted laborers and networks. Individuals from the partnership share encounters and team up on arrangements and systems to speed up the change away from coal.

The Global Energy Organization (IEA) assumes a critical part in working with worldwide coordinated effort on energy-related issues, including coal. The IEA's

Spotless Coal Community, laid out in 1975, centers around advancing perfect and productive coal advancements. While perceiving the natural difficulties related with coal, the IEA attempts to propel advancements, for example, carbon catch, usage, and capacity (CCUS) to decrease emanations from coal-terminated power plants. The organization's endeavors feature the logical methodology of tending to the natural effect of coal while perceiving its proceeded with job in the worldwide energy blend.

Two-sided coordinated efforts between nations likewise add to tending to coal-related issues. For instance, the US and China, as two of the world's biggest coal purchasers, have participated in joint drives to propel clean energy advances. The U.S.-China Clean Energy Exploration Center (CERC) works with cooperative innovative work on clean energy, including the improvement of cutting edge coal advances with lower discharges.

The World Bank, as a significant global monetary establishment, assumes a basic part in molding worldwide energy strategies. The World Bank Gathering's Energy Area The executives Help Program (ESMAP) upholds nations in their endeavors to progress to cleaner energy sources. The World Bank has focused on getting rid of funding for upstream oil and gas after 2019 and is progressively zeroing in on supporting nations in their change away from coal through interests in environmentally friendly power and energy effectiveness.

Territorial joint efforts offer stages for tending to coal-related issues with an emphasis on shared difficulties and open doors. The European Association (EU) has been at the very front of provincial endeavors to progress away from coal. The EU's obligation to the European Green Arrangement incorporates systems for a feasible and fair change, perceiving the social and financial effects on coal-subordinate districts. The European Commission's Foundation on Coal Districts On the move fills in as a gathering for sharing prescribed procedures and supporting locales in their change to a low-carbon economy.

The African Association, through its Plan 2063 system, perceives the significance of feasible energy access for the landmass' turn of events. While coal stays an energy source in a few African nations, the African Association is effectively advancing environmentally friendly power and clean innovations to address both energy access and ecological supportability. Cooperative drives inside the African Association plan to use global help for clean energy ventures and limit building endeavors.

The Asia-Pacific Financial Participation (APEC) discussion gives a stage to part economies to team up on energy issues, including coal-related difficulties. The APEC Energy Working Gathering centers around progressing reasonable and clean energy arrangements, and part economies share encounters and best practices to work with the change away from coal. Perceiving the variety of energy needs in the district, APEC underscores the significance of a fair methodology that thinks about financial turn of events, energy security, and ecological manageability.

The Icy Board, made out of eight Cold countries, addresses interesting ecological difficulties confronting the area, including the effects of dark fossil fuel byproducts from coal and different sources. While the Icy is definitely not a huge customer of coal, it is impacted by lengthy reach transport of contaminations. Cooperative endeavors inside the Cold Committee mean to decrease dark fossil fuel byproducts and moderate the natural and environment influences on the Icy locale.

The Global Work Association (ILO) perceives the social components of progressing away from coal. As a component of the "Rules for a Simply Change towards Ecologically Supportable Economies and Social orders for All," the ILO underscores the significance of comprehensive and fair strategies that consider the prosperity of laborers and networks impacted by the shift away from coal. Worldwide coordinated efforts including the ILO center around creating techniques to guarantee that the change is socially and doesn't abandon weak networks.

China, as the world's biggest coal shopper and maker, assumes a pivotal part in forming worldwide endeavors to address coal-related issues. While China keeps on depending on coal for a huge piece of its energy needs, the nation has likewise made significant interests in environmentally friendly power and clean innovations.

China's cooperation in worldwide coordinated efforts, like the Controlling Past Coal Collusion and joint drives with the US, mirrors its acknowledgment of the need to offset financial improvement with ecological manageability.

India, one more significant coal buyer, faces the test of satisfying its developing energy need while tending to natural worries. The Worldwide Sun based Union (ISA), sent off by India and France, expects to advance sun oriented energy and decrease reliance on petroleum derivatives, including coal. India's Public Activity Plan on Environmental Change incorporates drives to increment energy productivity, upgrade sustainable power limit, and advance practical turn of events.

The Belt and Street Drive (BRI), drove by China, has brought up issues about the natural effect of foundation projects, including coal-terminated power plants, in partaking nations. The cooperative idea of the BRI gives an open door to global commitment to guarantee that activities line up with supportable advancement objectives and natural norms. Endeavors to incorporate clean energy arrangements into the BRI highlight the significance of tending to coal-related issues inside the more extensive setting of framework improvement.

The Unified Countries Climate Program (UNEP) adds to worldwide endeavors to address coal-related natural difficulties. UNEP's work incorporates drives zeroed in on air quality, contamination control, and changing to manageable energy sources. The UNEP-drove Environment and Clean Air Alliance (CCAC) addresses brief environment contaminations, including dark carbon from coal burning, stressing the requirement for quick activity to lessen these poisons' effect on the environment and general wellbeing.

While global coordinated efforts and arrangements offer a system for tending to coal-related issues, the viability of these drives depends on the responsibility and activities of individual countries. The intricacies of energy frameworks, monetary contemplations, and homegrown political needs impact the degree to which nations carry out and stick to worldwide responsibilities. The intentional idea of numerous arrangements highlights the significance of cultivating a feeling of divided liability and shared benefit between countries.

Challenges in worldwide coordinated efforts on coal-related issues incorporate different public interests, financial contemplations, and the international setting. Arranging arrangements that balance the requirement for a quick change away from coal with the monetary and social worries of coal-subordinate locales requires political artfulness. The potential for "carbon spillage," where discharges are essentially uprooted to locales with less rigid natural guidelines, highlights the significance of facilitated worldwide endeavors to address coal-related difficulties.

The adequacy of global joint efforts additionally relies upon vigorous observing, announcing, and confirmation instruments to follow progress and consider countries responsible.

Straightforwardness and data sharing are essential for building trust among taking an interest nations and it are met to guarantee that responsibilities. The job of worldwide associations, for example, the UNFCCC, in working with normal evaluations and surveys of nations' activities is fundamental for keeping up with the energy of cooperative endeavors.

6.3 Countries successfully implementing policies for a sustainable energy transition.

The basic of progressing to feasible energy frameworks has become progressively apparent notwithstanding environmental change, ecological corruption, and the limited idea of customary energy sources. A few nations have arisen as pioneers in executing strategies for a reasonable energy progress, exhibiting a pledge to diminishing fossil fuel byproducts, expanding the portion of environmentally friendly power in their energy blend, and cultivating development in clean advances. Inspecting the encounters of these nations gives significant experiences into the different strategy moves toward that can drive an effective and practical energy progress.

Germany stands apart as a pioneer in the worldwide quest for a reasonable energy change, frequently alluded to as the "Energiewende." Started in the mid 2000s, the Energiewende addresses a thorough arrangement of strategies pointed toward progressing from petroleum derivatives to sustainable power sources. Key components of Germany's methodology incorporate the development of sustainable power limit, energy effectiveness measures, and a pledge to getting rid of atomic power.

A sign of Germany's energy change is the Environmentally friendly power Sources Act (EEG), sanctioned in 2000 and hence reexamined. The EEG presented a feed-in duty framework, giving ensured installments to environmentally friendly power makers, in this manner boosting the development of environmentally friendly power limit. This strategy has been instrumental in cultivating the sending of sun oriented, wind, and biomass innovations, making Germany a worldwide forerunner in environmentally friendly power limit.

The Energiewende likewise underlines energy productivity measures to decrease generally energy utilization. Germany has carried out projects to further develop energy productivity in structures, transportation, and modern cycles. The objective is to improve energy use across areas, guaranteeing that the change to sustainable power is joined by endeavors to limit energy wastage.

Germany's obligation to progressively eliminating atomic power originates from both natural and wellbeing contemplations. The choice to speed up the conclusion of thermal energy stations lines up with the more extensive objective of decreasing reliance on non-sustainable power sources. Nonetheless, this part of the Energiewende has confronted difficulties, including worries about energy security and the requirement for elective sources during the change.

Denmark has likewise arisen as a worldwide forerunner in manageable energy progress, especially in wind energy. The Danish government has set aggressive focuses for getting rid of non-renewable energy sources and accomplishing a carbon-unbiased economy by 2050. Wind energy assumes a focal part in Denmark's sustainable power procedure, with wind turbines providing a critical piece of the nation's power.

Denmark's outcome in wind energy is credited to a mix of strategy measures, innovative work drives, and public help. The nation executed feed-in levies, charge impetuses, and good administrative systems to energize interest in wind power. Also, cooperative endeavors between the public authority, industry, and research establishments have driven development in wind turbine innovation, making Danish organizations worldwide forerunners in the breeze energy area.

The combination of wind energy into Denmark's power lattice has been a key concentration, with headways in network the board and energy stockpiling. Denmark's experience highlights the significance of a comprehensive methodology that tends to the age of environmentally friendly power as well as the difficulties of coordinating that energy into existing framework.

Sweden has taken critical steps in accomplishing a supportable energy progress, with a solid accentuation on environmentally friendly power and a guarantee to decarbonization. The Swedish government has set aggressive targets, holding back nothing energy by 2040 and net-zero fossil fuel byproducts by 2045. Sweden's prosperity lies in its assorted and coordinated way to deal with maintainable energy.

Hydropower has generally assumed a significant part in Sweden's energy blend, giving a steady and sustainable wellspring of power. The nation has likewise put resources into other sustainable power sources, including wind and biomass. Sweden's methodology includes a blend of strategies, like feed-in levies, charge motivating forces, and innovative work subsidizing, to animate the development of sustainable power limit.

Past the attention on renewables, Sweden has carried out measures to further develop energy productivity across areas. Severe building regulations, energy effectiveness guidelines for machines, and drives elevating feasible transportation add to Sweden's general methodology for lessening energy utilization and fossil fuel byproducts.

Costa Rica has acquired global approval for its obligation to sustainable power and natural protection. The nation has set aggressive targets, planning to become carbon-nonpartisan by 2050. Costa Rica's progress in maintainable energy change is described by its dependence on sustainable sources, especially hydropower, wind, geothermal, and sun based energy.

Hydropower has generally been a critical supporter of Costa Rica's power age, utilizing the country's plentiful water assets. Lately, Costa Rica has expanded its environmentally friendly power portfolio by putting resources into wind and sun based projects. The public authority has executed approaches to draw in confidential interest in sustainable power, offering motivating forces and laying out clear administrative systems.

One remarkable accomplishment is Costa Rica's capacity to depend completely on environmentally friendly power for expanded periods. The nation has encountered times of continuous days where power was created exclusively from inexhaustible sources, displaying the possibility of a change away from petroleum derivatives.

Iceland's energy change is set apart by its bountiful geothermal assets, which have turned into the essential wellspring of power and warming in the country. Geothermal power plants bridle the World's inner intensity to produce power and give warming to homes and enterprises. Iceland's obligation to geothermal energy lines up with its objective of accomplishing energy freedom and diminishing dependence on petroleum products.

The usage of geothermal energy in Iceland is a consequence of long haul arranging, exploration, and interest in geothermal innovation. The public authority has effectively upheld the improvement of geothermal activities through monetary impetuses and coordinated effort with private and public partners. Iceland's progress in geothermal energy fills in as a model for nations with comparable land conditions.

Notwithstanding geothermal, Iceland has additionally put resources into hydroelectric power, further differentiating its environmentally friendly power portfolio. The nation's experience features the significance of utilizing one of a kind local assets and creating particular skill to drive an economical energy change.

Norway's energy change is eminent for its emphasis on hydropower and zap of the transportation area. Hydropower has been a foundation of Norway's energy creation for a really long time, giving a perfect and sustainable wellspring of power. The nation's geology and bountiful water assets make it appropriate for hydropower age.

Norway's obligation to feasible transportation includes an essential shift towards electric vehicles (EVs). The public authority has executed a scope of motivating forces, including tax reductions, cost exceptions, and public charging foundation improvement, to energize the reception of EVs. Norway has reliably positioned among the worldwide forerunners in EV piece of the pie, displaying the effect of steady strategies on the progress to cleaner transportation.

The jolt of the transportation area lines up with Norway's more extensive objective of decreasing fossil fuel byproducts across all areas. The incorporation of environmentally friendly power and supportable transportation mirrors a far reaching way to deal with accomplishing a low-carbon and economical future.

Uruguay's effective energy change is described by its obligation to renewables and endeavors to expand its energy blend. The nation has gained significant headway in a somewhat brief time frame, with an emphasis on wind and sunlight based energy. Uruguay's change is driven by a mix of strategy measures, market elements, and public help.

The public authority of Uruguay executed a serious closeout framework for sustainable power projects, drawing in confidential venture and driving down the expense of wind and sun oriented energy. The progress of these barterings has added to Uruguay's capacity to create a critical piece of its power from sustainable sources. The nation has likewise put resources into framework foundation to oblige the changeability of wind and sunlight based power.

Public help for environmentally friendly power and ecological manageability plays had a critical impact in Uruguay's energy change. The obligation to renewables is implanted in the public energy strategy, mirroring an agreement among residents, policymakers, and industry partners.

South Korea has embraced a diverse way to deal with accomplish a practical energy progress, stressing environmentally friendly power, energy productivity, and green innovation development. The South Korean government has set aggressive targets, including expanding the portion of environmentally friendly power, lessening fossil fuel byproducts, and cultivating a green economy.

Sustainable power sources, especially sun oriented and wind, have acquired conspicuousness in South Korea's energy blend. The public authority has executed feed-in duties, monetary motivations, and administrative systems to advance the development of environmentally friendly power limit. South Korea's obligation to renewables lines up with its more extensive technique to decrease reliance on petroleum derivatives and relieve the ecological effect of energy utilization.

The basic of accomplishing a feasible energy progress has turned into a worldwide need, driven by the acknowledgment of the ecological, monetary, and social difficulties related with conventional energy sources. The progress to manageable energy includes a key shift from petroleum derivatives to environmentally friendly power, combined with enhancements in energy effectiveness, creative advancements, and changes in energy utilization designs. This progress is fundamental for moderating environmental change, diminishing contamination, improving energy security, and encouraging long haul financial turn of events. Looking at the key components, difficulties, and achievement factors with regards to a supportable energy change gives significant bits of knowledge into the intricate and interconnected nature of this worldwide undertaking.

Environmentally friendly power as the Foundation:

Integral to the idea of a practical energy progress is the expanded use of sustainable power sources. These sources incorporate sunlight based, wind, hydropower, geothermal, and biomass, which offer the upside of being bountiful, clean, and fit for renewing themselves over the long haul.

Bridling the force of these renewables is basic for lessening fossil fuel byproducts, as the ignition of petroleum derivatives is a significant supporter of the ozone harming substances liable for environmental change. The organization of sustainable power advancements not just addresses the natural effects of energy creation yet additionally upgrades energy versatility by expanding the energy blend.

Strategy Systems and Administrative Help:

The outcome of a feasible energy change depends on the definition and execution of strong strategy structures and administrative measures. States assume a focal part in establishing an empowering climate for the change by setting clear targets, giving motivations, and laying out administrative systems that help sustainable power improvement. Feed-in duties, tax reductions, sustainable portfolio norms, and other strategy instruments are regularly utilized to animate interest in environmentally friendly power projects. A steady and unsurprising strategy climate is vital for drawing in private-area venture and encouraging development in clean energy advances.

Interest in Innovative work:

Development is a vital driver of an economical energy progress. Interest in innovative work (Research and development) is fundamental for propelling advancements, further developing effectiveness, and diminishing the expenses of sustainable power. State run administrations, industry, and the scholarly community team up to finance Research and development drives pointed toward creating state of the art arrangements, capacity innovations, and network joining systems. Leap forwards in energy capacity, shrewd networks, and materials science add to the versatility and unwavering quality of sustainable power sources, making them progressively aggressive with conventional petroleum products.

Innovative Headways:

Headways in environmentally friendly power advancements assume a urgent part in the progress to maintainable energy. Sun oriented photovoltaic (PV) boards, wind turbines, high level geothermal frameworks, and further developed energy capacity innovations have gone through critical headways, prompting expanded productivity and decreased costs. The adaptability of these advances considers their organization across different scales — from limited scope disseminated age to huge utility-scale projects. The constant quest for mechanical development guarantees that sustainable power turns into a suitable and practical option in contrast to ordinary energy sources.

Decentralization and Conveyed Energy Assets:

A supportable energy progress is described by a shift toward decentralized energy frameworks that influence circulated energy assets (DERs). Dispersed age, energy capacity, and request side administration empower a stronger and adaptable energy foundation. Decentralization decreases transmission and appropriation misfortunes, upgrades lattice unwavering quality, and engages neighborhood networks to create, store, and deal with their energy.

The ascent of housetop sun powered establishments, local area claimed sustainable ventures, and microgrids epitomizes the push toward a more decentralized and democratized energy scene.

Energy Proficiency Measures:

Further developing energy proficiency is a crucial part of a feasible energy progress. Energy productivity measures lessen generally speaking energy utilization, decline ozone harming substance discharges, and upgrade the monetary seriousness of businesses. States, organizations, and people execute energy-proficient innovations, practices, and building principles to improve energy use across areas. Energy proficiency measures incorporate overhauling building protection, taking on energy-productive apparatuses, and carrying out modern cycles that limit energy squander. The quest for energy productivity adds to the general manageability of the energy framework.

Jolt of Transportation:

The jolt of the transportation area is a pivotal part of a practical energy progress. Moving from gas powered motor vehicles to electric vehicles (EVs) lessens reliance on non-renewable energy sources and brings down fossil fuel byproducts. State run administrations overall are executing strategies to boost the reception of EVs, including sponsorships, tax reductions, and the advancement of charging framework. The combination of environmentally friendly power sources with electric transportation makes collaborations that add to a cleaner and more feasible portability biological system.

Round Economy and Economical Practices:

A reasonable energy change stretches out past the energy area to incorporate more extensive standards of manageability. The idea of a round economy underscores lessening waste, reusing materials, and reusing assets. Coordinating round economy standards into the energy progress includes maintainable practices like reusing parts of sustainable power frameworks, limiting natural effects, and guaranteeing capable obtaining of materials. This comprehensive methodology lines up with the more extensive objective of accomplishing manageability across monetary, social, and natural aspects.

Social and Monetary Contemplations:

An effective practical energy change perceives the social and monetary aspects inborn simultaneously. Simply Progress strategies plan to address the likely friendly and financial effects on networks and laborers impacted by the getting rid of conventional energy sources. These arrangements incorporate retraining programs, financial enhancement drives, and social security nets to help people and locales going through changes in the energy scene. Adjusting the financial goals of occupation creation and industry development with the social elements of value and inclusivity is fundamental for building broad help for the change.

Worldwide Cooperation and Associations:

Tending to the difficulties of a feasible energy progress requires worldwide joint effort and organizations. Worldwide participation permits nations to share best practices, advancements, and monetary assets. Drives like the Paris Arrangement unite countries to all in all address environmental change and focus on lessening ozone depleting substance emanations. Multilateral associations, research foundations, and industry coordinated efforts work with the trading of information and mastery, speeding up the worldwide change to practical energy.

Challenges in the Change:

Regardless of the huge advancement in reasonable energy change, different difficulties persevere. One of the essential difficulties is the discontinuity and inconstancy of sustainable power sources, for example, sun based and wind. Tending to this challenge requires headways in energy capacity advances, lattice adaptability, and request reaction measures to supply guarantee a dependable and stable energy.

The monetary parts of the progress present another obstacle. While the expenses of sustainable power innovations have diminished, forthright speculation stays a boundary for certain nations and organizations. Admittance to reasonable funding and the advancement of creative supporting systems are vital for beating this test and assembling the essential capital for huge scope sustainable power projects.

The current energy framework and the inactivity related with customary energy frameworks present difficulties to the fast joining of practical other options. Progressing from unified, non-renewable energy source based frameworks to decentralized, sustainable based frameworks requires massive changes in foundation, administrative structures, and plans of action.

Social acknowledgment and mindfulness likewise assume a basic part. Connecting with networks, addressing concerns connected with energy change, and guaranteeing that the advantages are evenhandedly dispersed are fundamental for building public help. Protection from change, particularly in areas reliant upon conventional energy enterprises, highlights the significance of successful correspondence and comprehensive dynamic cycles.

The international scene presents intricacies in the progress, as countries explore issues connected with energy security, asset contest, and international strains.

Guaranteeing an equitable and fair progress on a worldwide scale requires discretionary endeavors, joint effort, and a common obligation to tending to normal difficulties.

Chapter 7

The Dawn of Electricity

The beginning of power denoted a groundbreaking period in mankind's set of experiences, introducing another time of development and progress that would reshape social orders and reclassify the manner in which individuals lived, worked, and conveyed. It was a time of significant change, described by the outfitting of a power that had extended stayed a baffling and undiscovered force of nature.

The narrative of power starts with the spearheading work of researchers and designers who tried to grasp the basic standards of this confounding power. One of the critical figures in this investigation was Benjamin Franklin, whose examinations with lightning and his well known kite-flying trial in 1752 gave pivotal bits of knowledge into the idea of power. Franklin's work established the groundwork for additional examinations, starting interest and interest among researchers across the globe.

In the many years that followed Franklin's trials, researchers like Alessandro Volta, André-Marie Ampère, and Michael Faraday made earth shattering disclosures that developed comprehension we might interpret power. Volta's creation of the voltaic heap in 1800 denoted a huge achievement, giving a solid wellspring of constant electric flow. Ampère's work on electromagnetism and Faraday's trials with electromagnetic enlistment made ready for the improvement of electric generators and engines.

As the hypothetical basis for power was being laid out, useful applications started to arise. In the mid nineteenth hundred years, the message became perhaps the earliest significant innovation to outfit the force of power for correspondence. The creation of the message by Samuel Morse during the 1830s altered significant distance correspondence, empowering messages to be communicated quickly across huge spans utilizing electrical signs.

The mid-nineteenth century saw the advancement of the electric message organization, associating urban communities and locales in manners recently thought unimaginable. The immediate transmission of data over significant distances had significant ramifications for business, strategy, and regular daily existence. The message sped up the speed of correspondence as well as laid the preparation for the worldwide interconnectedness that would characterize the advanced world.

The following jump in the charge venture accompanied the creation of the electric light. Thomas Edison, the productive American creator, and his group worked resolutely to foster a commonsense and industrially reasonable glowing light. In 1879, Edison effectively showed his electric lighting framework, enlightening Menlo Park, New Jersey, with electric light interestingly. This cutting edge denoted the start of far reaching zap, as urban areas and towns steadily supplanted gas lights with electric lighting.

The reception of electric lighting significantly affected society, changing the manner in which individuals lived and worked. Night turned into an expansion of the day, and organizations could work longer hours. Streetlamps enlightened metropolitan scenes, upgrading wellbeing and security. The jolt of homes achieved an upset in home-grown life, supplanting candles and oil lights with the comfort and effectiveness of electric light.

The developing interest for power prompted the advancement of force circulation frameworks. George Westinghouse, a critical figure in the charge development, supported the utilization of exchanging current (AC) for significant distance power transmission. Westinghouse's vision, joined with the specialized mastery of designers like Nikola Tesla, prompted the development of force establishes that created power and dispersed it over progressively broad organizations.

The zap of industry was one more critical result of the ascent of power. Industrial facilities and assembling processes that had depended on physical work and steam power were currently ready to tackle the productivity and flexibility of electric engines. Creation turned out to be more smoothed out and proficient, prompting expanded yield and monetary development. The electric engine turned into a pervasive part in hardware, driving everything from transport lines to sequential construction systems.

The twentieth century saw a fast extension of electrical framework, driven by the rising interest for power in all parts of life. Power plants, at first filled by coal and later by oil and petroleum gas, multiplied to meet the developing power needs of a prospering modern and private scene.

The improvement of hydroelectric dams additionally expanded the wellsprings of power age, outfitting the force of streaming water to create perfect and sustainable power.

The creation of the radio and the ensuing improvement of the telecom business were crucial crossroads throughout the entire existence of power. Guglielmo Marconi's fruitful showing of remote telecommunication in the late nineteenth century established the groundwork for the remote transmission of sound signs. The radio turned into a strong vehicle for diversion, news, and correspondence, interfacing individuals across tremendous distances and molding mainstream society.

The mid-twentieth century saw the coming of TV, adding a visual aspect to the communicated content. TVs turned into a staple in families, giving a window to the world and forming public talk. The combination of power and gadgets led to the hardware business, making ready for advancements like semiconductors, incorporated circuits, and chip.

Space investigation turned into one more boondocks for the utilization of power and hardware. The improvement of satellites, space tests, and monitored shuttle depended intensely on cutting edge electrical frameworks for correspondence, route, and control. The Apollo missions, finishing in the notable moon arriving in 1969, displayed the capacities of electrical and electronic advances in the most difficult conditions.

The computerized transformation of the late twentieth century denoted a change in perspective in the connection among power and data. The creation of the PC and the resulting improvement of the web changed the manner in which individuals got to, shared, and handled data. Silicon chips and semiconductor innovations became basic to the working of PCs and electronic gadgets, energizing a period of quick mechanical headway.

The expansion of PCs, cell phones, and the web had extensive ramifications for correspondence, business, and social association. The world turned out to be progressively interconnected, and data turned out to be more available than any other time. The ascent of the data age was portrayed by the democratization of information and the strengthening of people through advanced innovations.

Sustainable power sources acquired unmistakable quality in the late twentieth and mid 21st hundreds of years as worries about ecological manageability and environmental change developed. Sunlight powered chargers and wind turbines became practical options in contrast to conventional petroleum product based power age, offering spotless and sustainable wellsprings of power. Propels in energy capacity advances additionally improved the unwavering quality and effectiveness of environmentally friendly power frameworks.

Savvy matrices and energy-productive advancements arose as basic parts of current electrical foundation. The mix of computerized sensors, robotization, and progressed investigation empowered more effective administration of power circulation, lessening waste and enhancing energy utilization. Savvy homes outfitted with interconnected gadgets permitted inhabitants to screen and control their energy use, adding to a more practical and harmless to the ecosystem way of life.

The auto business went through a progressive change with the coming of electric vehicles (EVs). Electric vehicles, controlled by battery-powered batteries, offered a cleaner and more economical option in contrast to customary gas powered motor vehicles. The advancement of high-limit batteries and upgrades in charging framework sped up the reception of electric vehicles, preparing for a future with diminished dependence on petroleum derivatives.

As the 21st century unfurled, the difficulties and valuable open doors related with power kept on molding the direction of human progress. The requirement for maintainable and strong energy frameworks turned out to be progressively critical as the effects of environmental change turned out to be more evident. Developments in energy capacity, network advancements, and environmentally friendly power sources assumed a vital part in tending to these difficulties and building a more supportable future.

The joining of computerized reasoning (artificial intelligence) into electrical frameworks opened new outskirts in robotization, proficiency, and direction. Simulated intelligence calculations were conveyed to streamline energy matrices, anticipate request examples, and upgrade the unwavering quality of electrical framework. The marriage of artificial intelligence and power made ready for brilliant urban areas, where interconnected innovations and information driven arrangements were bridled to work on metropolitan residing.

The charge of transportation stretched out past individual vehicles to incorporate public travel and even aeronautics. Electric trains, transports, and planes became suitable options in contrast to their petroleum product partners, adding to endeavors to diminish ozone harming substance discharges and battle environmental change. The vision of a completely energized and supportable transportation framework picked up speed as mechanical progressions and ecological objectives merged.

In the domain of medical services, power kept on assuming a fundamental part in demonstrative and remedial applications. Clinical imaging innovations, for example, X-beams, attractive reverberation imaging (X-ray), and registered tomography (CT) examines, depended on the age and control of electrical signs to give point by point experiences into the human body. Electrically fueled clinical gadgets and hardware became irreplaceable devices for conclusion, treatment, and exploration.

The union of science and power led to the field of bioelectronics, where electronic gadgets interacted with organic frameworks to screen and tweak physiological cycles. Implantable gadgets, like pacemakers and neurostimulators, exemplified the capability of bioelectronics in further developing wellbeing and personal satisfaction. The connection point among innovation and science opened additional opportunities for tending to clinical difficulties and propelling the outskirts of medical care.

As the world turned out to be progressively dependent on power, the significance of network safety in safeguarding basic framework turned into a squeezing concern. The interconnectedness of electrical frameworks, combined with the multiplication of computerized innovations, made new weaknesses that could be taken advantage of by malevolent entertainers. States, organizations, and people the same needed to put resources into powerful online protection measures to shield against digital dangers and guarantee the dependability of electrical organizations.

The quest for combination energy, an interaction that recreates the sun's power, turned into a point of convergence of logical examination and energy development. If effectively saddled, combination energy could give an essentially boundless and clean wellspring of force. In any case, the specialized difficulties and gigantic intricacy of accomplishing controlled combination responses presented imposing impediments that analysts kept on wrestling with.

The cultural effect of power reached out past the domain of innovation and foundation. The democratization of data, worked with by the web, enabled people and networks to participate in new types of correspondence and activism. Social developments, energized by the network managed the cost of by computerized innovations, resolved issues going from basic freedoms to natural equity, molding an additional internationally cognizant and interconnected world.

The difficulties of overseeing and circulating power in a quickly developing scene required continuous exploration and advancement. Energy capacity advances, like high level batteries and supercapacitors, assumed a pivotal part in tending to the irregularity of environmentally friendly power sources and upgrading the solidness of electrical lattices. Materials science and nanotechnology added to the improvement of additional effective and maintainable electrical parts.

The push for energy proficiency and maintainability prompted the improvement of green structure advances and brilliant foundation. Structures furnished with energy-proficient frameworks, environmentally friendly power sources, and savvy controls became models of practical metropolitan turn of events. The idea of the shrewd city advanced, consolidating standards of natural stewardship, asset productivity, and personal satisfaction.

The worldwide progress to sustainable power sources picked up speed, driven by a developing familiarity with the natural effects of petroleum product ignition. Worldwide endeavors to lessen fossil fuel byproducts and relieve environmental change provoked the reception of sustainable power targets and the advancement of approaches to boost clean energy speculations. The Paris Understanding, endorsed in 2015, denoted a huge achievement in the worldwide obligation to addressing environmental change and progressing to a low-carbon future.

The job of power in addressing worldwide difficulties stretched out to issues of social value and access. The computerized partition, the hole between the individuals who approach data and correspondence innovations and the people who don't, turned into a point of convergence for policymakers and backers. Endeavors were made to connect this separation, guaranteeing that the advantages of power and computerized advancements were open to all, independent of geological area or financial status.

In the domain of training, power worked with new methods of learning and information scattering. Online schooling stages, intelligent learning instruments, and computerized assets changed customary instructive ideal models, offering a more customized and open way to deal with learning. The jolt of instruction turned into an impetus for deep rooted getting the hang of, engaging people to obtain new abilities and information all through their lives.

The groundbreaking effect of power on society was joined by moral contemplations and cultural difficulties. The dependable turn of events and utilization of arising advances, for example, simulated intelligence and biotechnology, required cautious thought of moral standards, protection concerns, and likely cultural ramifications. The requirement for moral structures and administrative shields became principal to guarantee that the advantages of power were evenhandedly circulated and that the potential dangers were relieved.

The developing scene of energy international relations molded worldwide relations and key contemplations. The journey for energy security, combined with the international ramifications of energy assets, affected conciliatory relations and international elements. Nations looked to expand their energy sources, lessen reliance on petroleum products, and take part in coordinated efforts to address shared energy challenges.

The vision of a maintainable and energized future incited a reconsidering of metropolitan spaces and transportation frameworks. Brilliant and economical urban communities consolidated environmentally friendly power sources, savvy foundation, and productive transportation organizations. Electric vehicles, controlled by clean energy, became fundamental to the metropolitan versatility biological system, diminishing air contamination and relieving the natural effect of transportation.

The extraordinary excursion of power from its initial logical investigation to a pervasive and key power in current life highlighted the limit of human creativity to bridle and shape the powers of nature. The tale of power was one of development, steadiness, and cooperation, as people and networks across the globe cooperated to open the capability of this crucial power.

As the world kept on wrestling with complex difficulties, from environmental change to social imbalance, the job of power stayed vital to imagining and constructing an additional economical and evenhanded future. The illustrations gained from the historical backdrop of power filled in as a directing light, stressing the significance of dependable stewardship, moral contemplations, and an aggregate obligation to saddling the force of power to help mankind. The beginning of power, when a far off and secretive skyline, had turned into a signal enlightening the way to a future where the powers of nature and human development joined to ultimately benefit all.

7.1 Highlight success stories and breakthroughs in transitioning from coal to electricity.

The progress from coal to power addresses a great change in the energy scene, driven by a worldwide basic to decrease natural effect, battle environmental change, and embrace manageable energy sources. This progress has been set apart by various examples of overcoming adversity and forward leaps, exhibiting the potential for positive change when countries, businesses, and networks focus on a cleaner and more manageable energy future.

One of the outstanding examples of overcoming adversity in the progress from coal to power is exemplified by Germany's Energiewende, or energy change. Germany, a country with a generally solid dependence on coal, left on an aggressive excursion to build the portion of sustainable power in its power blend while eliminating coal-terminated power plants. The Energiewende, sent off in the mid 2000s, expected to accomplish a more feasible, low-carbon energy framework.

One critical forward leap in Germany's progress was the quick extension of environmentally friendly power limit, especially in wind and sun based. The nation executed strong strategies and motivating forces to energize the organization of environmentally friendly power advancements, prompting a significant expansion in wind ranches and sun based establishments. By 2020, sustainable power represented more than 40% of Germany's power age, denoting a critical achievement in the decrease of coal reliance.

One more example of overcoming adversity unfurled in the Unified Realm, where the change away from coal was advanced rapidly through a blend of strategy measures, market elements, and mechanical progressions. The UK's obligation to transitioning away from unabated coal-terminated power plants by 2024 prepared for expanded interest in cleaner energy choices. The downfall of coal was supplemented by a flood in flammable gas and environmentally friendly power limit, including seaward wind cultivates that turned into a conspicuous element of the UK's energy scene.

The conclusion of coal mineshafts and coal-terminated power plants introduced a perplexing test, especially in districts with a well established coal mining custom. Notwithstanding, fruitful models arose where networks progressed to new financial exercises and embraced open doors in the developing sustainable power area. Preparing projects and drives pointed toward reskilling the labor force assumed a pivotal part in supporting the progress, guaranteeing that the impacted networks could take part in the arising clean energy economy.

China, as the world's biggest customer and maker of coal, set out on a complex way to deal with change from coal to power. The nation confronted extreme air contamination challenges, basically determined by coal ignition, and perceived the need to move towards cleaner and more manageable energy sources. China turned into a worldwide forerunner in sustainable power sending, putting vigorously in sun based and wind power.

A prominent forward leap in China's progress was the quick development of its sun powered photovoltaic (PV) industry. The nation arose as the world's biggest maker and buyer of sunlight powered chargers, utilizing economies of scale to drive down the expense of sun based energy. The sending of enormous scope sunlight based ranches and the combination of appropriated sun powered establishments on roofs contributed fundamentally to decreasing coal reliance and alleviating natural effects.

India, confronting a double test of energy access and ecological manageability, sought after an aggressive plan to grow power access while diminishing dependence on coal. The nation carried out the Public Sunlight based Mission, meaning to accomplish 100 gigawatts (GW) of sun powered limit by 2022. India's obligation to environmentally friendly power stretched out past sun based, with huge interests in wind power and hydroelectric ventures.

One of the examples of overcoming adversity in India's change was the province of Gujarat, which embraced sun based energy with excitement and key preparation. The improvement of the Charanka Sun oriented Park, quite possibly of the biggest sun based park in Asia, exhibited the potential for enormous scope sun powered reconciliation. The recreation area united different sunlight based projects, showing the plausibility of decentralized environmentally friendly power age and giving an outline to different locales to follow.

In the US, the change from coal to power unfurled unevenly across various states and locales. A few states focused on aggressive environmentally friendly power targets and embraced strategies to boost the improvement of clean energy projects. California, for instance, turned into a pioneer in advancing sun based energy and energy proficiency measures, driving the manner in which in lessening coal reliance and ozone harming substance emanations.

The ascent of gaseous petrol assumed a huge part in the U.S. progress, offering a cleaner choice to coal for power age. The improvement of pressure driven cracking, or deep earth drilling, empowered the extraction of plentiful flammable gas holds, prompting a change in the energy blend. The expanded utilization of flammable gas in power plants added to a decrease in fossil fuel byproducts and gave an extension toward a more maintainable energy future.

Mechanical leap forwards in energy capacity have been instrumental in tending to the discontinuity of sustainable power sources, like breeze and sun oriented. Battery capacity innovations, including lithium-particle batteries, have encountered critical headways, making it conceivable to store overabundance energy created during times of high sustainable result and delivery it when request is high. These developments improve framework security and empower a smoother joining of environmentally friendly power into the power framework.

The charge of transportation arose as a vital system to lessen the carbon impression related with the transportation area, which vigorously depends on petroleum products. Electric vehicles (EVs) built up forward movement worldwide, with progressions in battery innovation prompting expanded driving reach and quicker charging abilities. States and organizations boosted the reception of EVs through endowments, tax breaks, and the advancement of charging framework.

Norway stands apart as an example of overcoming adversity in the jolt of transportation. The nation executed approaches to support the reception of electric vehicles, including charge motivations, cost exceptions, and devoted paths for EVs. Thus, Norway turned into a worldwide forerunner in electric vehicle reception, with electric vehicles comprising a huge portion of new vehicle deals. The progress of Norway's methodology showed the viability of strong strategies in speeding up the change to electric transportation.

Developments in lattice innovations and savvy matrices play had a critical impact in enhancing the coordination of environmentally friendly power into the power network. Shrewd lattices influence computerized innovations, sensors, and progressed investigation to improve the productivity and unwavering quality of power dispersion. These frameworks empower ongoing observing, request reaction, and the consistent combination of different energy sources, adding to a stronger and adaptable power foundation.

Microgrids, one more leap forward in power circulation, offer limited and decentralized answers for power age and conveyance. These more limited size, independent energy frameworks can work autonomously or related to the primary matrix, giving versatility notwithstanding interruptions and supporting local area level energy independence. Microgrids have demonstrated especially important in remote or off-matrix regions, where conventional framework foundation might be trying to execute.

The responsibility of global partnerships to progress to environmentally friendly power has been a distinct advantage in the more extensive shift from coal to power. Significant organizations across enterprises, including innovation, retail, and assembling, have laid out aggressive objectives to drive their activities with sustainable power. Corporate power buy arrangements (PPAs) for sustainable power projects have turned into a main impetus in the extension of clean energy limit.

Google, for example, accomplished a huge achievement by turning into the primary significant organization to counterbalance its whole functional carbon impression through the acquisition of environmentally friendly power. The organization focused on accomplishing 100 percent sustainable power for its worldwide tasks and has put resources into a different arrangement of environmentally friendly power projects, including wind and sunlight based ranches.

The job of global coordinated effort and drives couldn't possibly be more significant in that frame of mind from coal to power. The Paris Understanding, embraced in 2015, united countries from around the world in an aggregate obligation to restrict an Earth-wide temperature boost and diminish ozone depleting substance outflows. The arrangement set up for cooperative endeavors to speed up the change to clean energy and encourage reasonable turn of events.

Multilateral endeavors, like the Worldwide Sun based Collusion, have tried to activate worldwide help for the arrangement of sun powered energy. The partnership, sent off by India and France, means to work with collaboration among sun oriented rich nations to bridle sun based influence for both power age and different applications. By encouraging mechanical coordinated effort and sharing accepted procedures, such drives add to the worldwide force toward manageable energy changes.

The continuous progress from coal to power faces difficulties and intricacies, including the requirement for simply advances for networks subject to coal, tending to energy neediness, and guaranteeing the dependability of the power framework. Nonetheless, the examples of overcoming adversity and leap forwards framed here highlight the exceptional headway that has been made in reshaping the energy scene.

7.2 Showcase communities and industries thriving on clean energy solutions.

Networks and businesses overall are going through groundbreaking movements, embracing clean energy answers for address ecological worries, lessen carbon impressions, and assemble more reasonable and versatile fates. These examples of overcoming adversity feature the positive effect of embracing clean energy across different areas, displaying how networks and businesses can flourish while adding to a greener and more maintainable planet.

In Scandinavia, Denmark stands apart as a trailblazer in wind energy. The nation has effectively bridled its breezy waterfront conditions to turn into a worldwide forerunner in wind power age. Danish people group, both metropolitan and country, have embraced breeze ranches as a perfect and effective wellspring of power. One surprising model is the seaward wind ranch Middelgrunden, situated close to Copenhagen. This people group possessed breeze ranch comprises of 20 turbines and fills in as a model for maintainable energy creation, with neighborhood occupants effectively associated with its turn of events and activity.

In Germany, the idea of energy cooperatives has acquired noticeable quality, empowering networks to effectively take part in and benefit from the perfect energy change. These cooperatives, frequently claimed by nearby occupants, put resources into environmentally friendly power undertakings like breeze and sunlight based ranches. The residents of Schönau, an unassuming community in Germany, laid out one of the main energy cooperatives because of worries about atomic power. Today, the Schönau agreeable produces clean energy as well as adds to neighborhood financial turn of events and local area prosperity.

California, a pioneer in clean energy drives, has seen the development of flourishing businesses and networks focused on manageability. The state's forceful sustainable power targets and arrangements have prodded advancement and interest in clean advancements. The city of Lancaster, for example, has turned into a sustainable power center, with a noteworthy exhibit of sun oriented establishments. Lancaster's obligation to clean energy has drawn in organizations, made positions, and situated the city as a model for others looking to progress to a maintainable energy future.

The island country of Iceland has utilized its bountiful geothermal assets to make an economical and strong energy framework. Geothermal power gives most of Iceland's power and is utilized for warming homes, nurseries, and, surprisingly, pools. The town of Husavik represents the positive effect of geothermal energy on neighborhood networks. With its geothermal region warming framework, Husavik appreciates minimal expense and harmless to the ecosystem warming, cultivating a great of life for its inhabitants.

China, a worldwide forerunner in sun oriented photovoltaic (PV) assembling and sending, has seen the fast development of enterprises and networks embracing sun based power. The city of Dezhou is prestigious as China's "Sunlight based City," showing a pledge to sun oriented energy for an enormous scope. Dezhou's horizon is decorated with sun oriented establishments, from housetop sun powered chargers to sun based water warmers. This coordinated exertion has diminished the city's carbon impression as well as situated Dezhou as a middle for sunlight based innovation innovative work.

In the Unified Bedouin Emirates (UAE), the city of Dubai has arisen as a brilliant illustration of how clean energy can change the energy scene as well as the economy. The Mohammed receptacle Rashid Al Maktoum Sun powered Park, quite possibly of the biggest sun oriented park on the planet, has situated Dubai as a worldwide center point for sun based energy. The recreation area's imaginative ventures, including concentrated sun oriented power and the world's tallest sun based tower, exhibit Dubai's obligation to manageability and its aspiration to differentiate its energy sources.

Enterprises overall are perceiving the financial and ecological advantages of changing to clean energy arrangements. The corporate area, specifically, plays had a pivotal impact in driving the reception of sustainable power.

Tech monsters like Google, Apple, and Microsoft have focused on driving their tasks with 100 percent environmentally friendly power. Google, for example, accomplished carbon impartiality in 2007 and has been coordinating its energy utilization with environmentally friendly power buys, driving the way for different organizations to stick to this same pattern.

The car business has gone through a huge change with the ascent of electric vehicles (EVs). Tesla, an American electric vehicle and clean energy organization, has led this transformation. Tesla's Gigafactories, decisively situated all over the planet, produce electric vehicles and batteries at an uncommon scale. The progress of Tesla has disturbed the auto business as well as sped up the worldwide shift toward electric transportation.

In the farming area, manageable practices and clean energy arrangements are building up momentum. The idea of "brilliant cultivating" coordinates sustainable power innovations to upgrade agrarian proficiency. Sun oriented controlled water system frameworks, for instance, offer a maintainable answer for water the board in horticulture. Ranchers in locales like India have embraced sun oriented siphons to flood their fields, lessening reliance on non-renewable energy sources and traditional matrix power.

The sea business, generally dependent on petroleum derivatives, is likewise embracing clean energy options. The world's most memorable completely electric and independent compartment transport, the Yara Birkeland, addresses a noteworthy improvement in the sea area. This zero-emanation vessel, furnished with electric drive and independent route frameworks, embodies the potential for clean energy to upset conventional enterprises and diminish the ecological effect of delivery.

In the domain of engineering and metropolitan preparation, supportable plan standards and clean energy arrangements are molding the urban communities representing things to come. Masdar City in Abu Dhabi, Joined Middle Easterner Emirates, fills in as a model for maintainable metropolitan turn of events. Planned with an emphasis on environmentally friendly power, energy proficiency, and shrewd foundation, Masdar City expects to be a carbon-nonpartisan and zero-squander metropolitan focus. The city features how imaginative metropolitan arranging can establish conditions that focus on clean energy and maintainability.

Microgrids, decentralized energy frameworks that can work freely or related to the principal lattice, are assuming a crucial part in guaranteeing energy access and flexibility in remote and off-matrix networks. The Indian province of Kerala has carried out a fruitful microgrid project that has zapped a few far off towns. These microgrids, frequently fueled by sunlight based energy, give a dependable and feasible wellspring of power, working on the personal satisfaction for inhabitants and cultivating monetary turn of events.

In the medical services area, clean energy arrangements add to both natural manageability and worked on persistent consideration. Medical clinics overall are taking on energy-proficient advances, environmentally friendly power sources, and imaginative structure plans to decrease their carbon impression. The Cleveland Facility in Ohio, USA, remains as a striking model, with its obligation to manageability reflected in LEED-ensured structures, energy-productive frameworks, and environmentally friendly power establishments.

Networks on the cutting edge of environmental change are progressively going to clean energy for of building flexibility and moderating the effects of an evolving environment. In the Pacific Island country of Samoa, sunlight based power has turned into a basic part of the country's endeavors to address energy security and environment versatility. The reception of sun powered energy in Samoa has decreased dependence on imported petroleum products, giving a solid and supportable wellspring of power for the island country.

The progress of clean energy arrangements isn't restricted to well-to-do or innovatively progressed locales. Rustic people group in non-industrial nations are likewise encountering the extraordinary effect of clean energy mediations. The Shoeless School, situated in India, has spearheaded the utilization of sun oriented ability to enable country networks. The school trains ladies from these networks to become sun oriented engineers, outfitting them with the abilities to introduce, keep up with, and fix sun based power frameworks. This grassroots methodology has carried power to distant towns as well as enabled ladies and catalyzed neighborhood monetary turn of events.

Seaward wind ranches address a boondocks of clean energy improvement, saddling the immense capability of wind energy over untamed oceans. The Unified Realm, with its broad shoreline, has turned into a forerunner in seaward wind power. The London Cluster, situated in the Thames Estuary, was the world's biggest seaward wind ranch at the hour of its finishing. Seaward wind ranches offer a versatile and harmless to the ecosystem answer for satisfy developing energy needs while limiting the effect ashore biological systems.

The idea of round economies, where assets are utilized effectively, squander is limited, and environmentally friendly power is focused on, is getting some forward momentum around the world. The city of Amsterdam in the Netherlands has embraced the standards of a roundabout economy. Drives, for example, the Amsterdam Brilliant City project center around supportable energy arrangements, savvy framework, and waste decrease. By reconsidering the conventional straight monetary model, Amsterdam embodies how urban areas can flourish while focusing on maintainability and clean energy.

In the scholastic domain, colleges and exploration organizations are driving the manner in which in executing clean energy arrangements and advancing supportability. Stanford College in California, USA, has focused on accomplishing net-zero ozone harming substance discharges and acquiring 100 percent of its power from sustainable sources.

Stanford's creative methodologies, remembering for grounds sun oriented establishments and energy effectiveness measures, act as a model for instructive foundations looking to incorporate clean energy into their tasks.

7.3 Future prospects and challenges in sustaining the dawn of electricity beyond coal.

As we stand at the edge representing things to come, the possibilities for supporting the beginning of power past coal are both promising and testing. The excursion from coal to cleaner and more feasible energy sources has denoted a critical change in outlook, yet the street ahead requests proceeded with development, vital preparation, and worldwide cooperation to guarantee a strong and maintainable energy future.

One of the key future possibilities lies in the fast progression of sustainable power advancements. Sun based and wind power, specifically, have seen significant improvement concerning productivity, cost-adequacy, and versatility. Progressing innovative work endeavors intend to upgrade the presentation of sun powered chargers, increment the proficiency of wind turbines, and investigate novel ways to deal with tackling environmentally friendly power. As these advances keep on developing, they hold the commitment of giving solid and cost-serious options in contrast to customary non-renewable energy sources.

Energy capacity innovations are basic to tending to the irregular idea of environmentally friendly power sources. Progresses in battery advancements, including lithium-particle batteries and arising options, are opening up additional opportunities for putting away and effectively using power created from sun oriented and wind sources. Forward leaps in energy capacity can possibly reform lattice the executives, empowering a more solid and tough energy foundation equipped for obliging a higher portion of environmentally friendly power.

The jolt of transportation is one more groundbreaking pattern with critical future possibilities. The developing reception of electric vehicles (EVs) presents a chance to diminish dependence on petroleum derivatives in the transportation area, a significant supporter of fossil fuel byproducts. Upgrades in battery innovation, charging framework, and government impetuses are driving the broad reception of EVs. What's in store holds the potential for additional development in electric portability, remembering headways for battery range, charging speed, and the mix of savvy advances.

Savvy lattice advancements are ready to assume a urgent part in upgrading the coordination of environmentally friendly power into the power network. The improvement of wise lattice frameworks, furnished with cutting edge sensors, correspondence organizations, and information investigation, empowers continuous observing and control of power dispersion. Brilliant lattices upgrade matrix unwavering quality, further develop energy productivity, and work with the consistent reconciliation of different energy sources, including environmentally friendly power.

As the world moves towards decentralized energy frameworks, savvy matrices will turn out to be progressively fundamental for overseeing mind boggling and dynamic energy organizations.

Computerized reasoning (simulated intelligence) is arising as an incredible asset for improving energy frameworks and tending to the difficulties related with environmentally friendly power mix. Simulated intelligence calculations can estimate energy interest, advance energy creation and utilization, and improve the productivity of matrix activities. AI applications in energy the executives hold the possibility to adjust to evolving conditions, working on the dependability and strength of the power network progressively. The collaboration among simulated intelligence and clean energy advancements presents open doors for more wise, versatile, and proficient energy frameworks.

The worldwide obligation to lessening fossil fuel byproducts and moderating environmental change is a main impetus behind the progress to cleaner energy sources. The Paris Understanding, endorsed by nations all over the planet, sets focuses for restricting worldwide temperature increments and speeding up the change to a low-carbon future. The rising consciousness of natural manageability and the basic to control environmental change give areas of strength for a to proceeded with endeavors to support the beginning of power past coal.

Worldwide joint effort is a critical consider tending to worldwide energy challenges and propelling clean energy arrangements. Cooperative drives, for example, joint examination projects, innovation sharing, and strategy coordination, can speed up the turn of events and organization of clean energy innovations on a worldwide scale.

The trading of best practices, information move, and shared help among countries are fundamental for making a strong and successful system for maintainable energy improvement.

The ascent of decentralized energy frameworks, including microgrids and dispersed energy assets, addresses a change in perspective in how power is produced and consumed. Future possibilities incorporate the far reaching reception of confined energy arrangements that engage networks, improve energy versatility, and lessen dependence on concentrated power plants. The adaptability and flexibility of decentralized frameworks offer a decentralized energy frameworks, including microgrids and dispersed energy assets, addresses a change in perspective in how power is created and consumed. Future possibilities incorporate the boundless reception of confined energy arrangements that engage networks, improve energy strength, and lessen dependence on concentrated power plants. The adaptability and flexibility of decentralized frameworks offer a versatile and manageable way to deal with meeting the different energy needs of networks, especially in remote or underserved regions.

The idea of energy a majority rule government, wherein networks have more noteworthy control and responsibility for energy assets, is picking up speed. Local area based environmentally friendly power projects, energy cooperatives, and participatory models enable neighborhood

occupants to effectively participate in and benefit from the perfect energy progress. Future possibilities incorporate the proceeded with development of energy a majority rules government, with networks becoming key partners in the preparation, improvement, and the board of clean energy projects.

Notwithstanding the promising future possibilities, supporting the beginning of power past coal represents a few difficulties that request cautious thought and proactive arrangements. One critical test is the requirement for thorough energy framework updates. The progress to sustainable power sources requires modernizing and adjusting existing network frameworks to oblige fluctuating energy inputs, coordinate energy stockpiling arrangements, and backing bidirectional energy streams. The forthright expenses and calculated intricacies related with overhauling framework can present difficulties, requiring key preparation and speculation.

Discontinuity and fluctuation of environmentally friendly power sources stay key difficulties that influence framework strength. The innate unusualness of daylight and wind requires successful energy stockpiling arrangements and framework the board systems to adjust organic market. Developments in energy capacity advancements, request reaction components, and matrix adaptability are fundamental for conquering these difficulties and guaranteeing a dependable and tough power foundation.

The progress away from coal likewise raises worries about the social and monetary effects on networks that have generally depended on coal mining and coal-terminated power plants. Simply progress strategies and projects are significant for supporting impacted networks, giving retraining potential open doors, and working with the expansion of neighborhood economies. Guaranteeing that the perfect energy progress is fair and comprehensive requires a proactive way to deal with address the expected disengagement and monetary changes in locales reliant upon the coal business.

While sustainable power sources are by and large considered cleaner than petroleum products, the natural effect of specific advancements and their related stockpile chains should be painstakingly made due. For example, the creation and removal of sunlight powered chargers, wind turbines, and energy stockpiling gadgets include asset extraction, fabricating cycles, and waste administration contemplations. An extensive way to deal with maintainable works on, including reusing, capable obtaining, and limiting natural impression, is imperative to guarantee the by and large ecological supportability of clean energy innovations.

The monetary parts of progressing to clean energy additionally present difficulties. Albeit the expenses of environmentally friendly power advancements have diminished altogether lately, the underlying capital ventures expected for enormous scope organization can be significant.

Beating monetary boundaries includes a mix of strong strategies, monetary motivators, and creative supporting systems. Legislatures, organizations, and monetary foundations should team up to make good circumstances for drawing in speculations and increasing clean energy projects.

Energy security stays a basic thought as the world changes from coal. The need to guarantee a steady and dependable energy supply, particularly during the change time frame, requires cautious preparation and coordination. Differentiating energy sources, fortifying energy stockpiling capacities, and improving matrix flexibility are fundamental parts of a complete energy security technique.

The international ramifications of the spotless energy change are additionally significant. As nations diminish their dependence on petroleum derivatives, customary energy exporters might encounter shifts in financial elements and international impact. The requirement for new international systems, participation on energy security, and discretionary methodologies to explore these progressions is significant for encouraging worldwide steadiness and coordinated effort.

Administrative systems and strategy solidness assume an essential part in molding the eventual fate of clean energy. Conflicting or questionable approaches can make hindrances to venture and ruin the development of clean energy markets. Clear, steady, and strong administrative conditions are fundamental to draw in confidential area ventures, empower long haul arranging, and encourage advancement in the perfect energy area.

www.ingramcontent.com/pod-product-compliance
Lightning Source LLC
LaVergne TN
LVHW050650200726
843506LV00010B/1449